A-Z of Clinical Chemistry

A-Z of Clinical Chemistry

A Guide for the Trainee

W. Hood

Principal Biochemist,
Tameside General Hospital,
Ashton-under-Lyne

MTP PRESS LIMITED
International Medical Publishers

Published in the UK by
MTP Press Limited
Falcon House
Lancaster, England

Softcover reprint of the hardcover 1st edition 1980

First published 1980

British Library Cataloguing in Publication Data

Hood, W
A–Z of clinical chemistry.
1. Chemistry, Clinical
I. Title
616.07'56 RB40
ISBN-13: 978-94-011-6662-1 e-ISBN-13: 978-94-011-6660-7
DOI: 10.1007/978-94-011-6660-7

Contents

Preface

The newcomer to the field of clinical chemistry is faced with the daunting prospect of understanding the ever increasing aspects of the subject: new techniques, tests, terminology, methods of diagnosing diseases and other advances which relate to clinical chemistry. The aim of this book is to provide basic information regarding all branches of the subject which the trainee will need to understand. The book should also provide a basis for answering many of the examination questions of clinical chemistry. It is therefore hoped that this book will prove useful to any person starting a career in clinical chemistry, be that person a laboratory scientific officer, graduate or trainee pathologist. Wherever possible, suggestions for further reading are given. Many subjects are so broad however that the reader is referred to the general list of analytical and clinical textbooks supplied at the end of the book.

My grateful thanks are extended to the three ladies who typed this manuscript: Christine Cliffe, Margaret Donnelly and Judith Hardy. I would like to thank Dr Brian Wisdom of Queens University, Belfast for his suggestions on the entry "Enzyme-immunoassay". Finally my thanks go to Mr Martin Lister of MTP Press for his help and encouragement in this venture.

W. H.
ASHTON-UNDER-LYNE
APRIL 1980

A

ABETALIPOPROTEINAEMIA

A rare hereditary disorder in which there is a complete absence of β-lipoprotein, pre-β-lipoprotein and chylomicrons. It presents clinically as ataxia and malabsorption with steatorrhoea. Thorny shaped erythrocytes (acanthocytes) are a feature of the disease.

Further reading: Herbert, P.N., Gotto, A.M. and Fredrickson, D.S. (1978). Familial lipoprotein deficiency (abetalipoproteinaemia, hypobetalipoproteinaemia, and Tangier disease). In Stanbury, J.B., Wyngaarden, J.B. and Fredrickson, D.S. (eds.) *The Metabolic Basis of Inherited Disease,* 4th Edn., p. 544. (New York: McGraw-Hill)

ABSORBANCE (OPTICAL DENSITY, EXTINCTION)

The amount of light absorbed by a solution. Beer's law states that the amount of light absorbed by a solution is proportional to its concentration. Lambert's law states that the amount of light absorbed by a solution is proportional to the length of the light path. A combination of these two laws gives the expression

$$A = Kct$$

where A is the absorbance (i.e. the amount of light absorbed),
K is a constant,

c is the concentration of the absorbing species and

t is the length of the light path

Since t, the length of the light path, is normally kept constant, it follows that the absorbance is proportional to the concentration of the substance being measured.

Further reading: General list of analytical textbooks

ACATALASAEMIA

A rare inborn error in which there is a deficiency of catalase in red blood cells and tissues. It is transmitted as an autosomal recessive. Patients with the condition can present with ulcerating lesions of the mouth. This is thought to be due to the harmful effects of hydrogen peroxide (produced by bacteria within the oral cavity) which would otherwise be broken down by the catalase.

Further reading: Aebi, H.E. and Wyss, S.R. (1978). Acatalasaemia. In Stanbury, J.B., Wyngaarden, J.B. and Fredrickson, D.S. (eds.) *The Metabolic Basis of Inherited Disease* 4th Edn. p. 1792 (New York: McGraw-Hill).

ACCURACY

The extent to which measurements of a substance differ from its true value.

See also: **precision**

ACETAZOLAMIDE

A diuretic drug sometimes used in the treatment of glaucoma. It has to be used with care because of the danger of acidosis. This is because it inhibits carbonic anhydrase. In the kidney this results in impaired bicarbonate reabsorption and an associated loss of sodium in the urine.

ACETEST

Reagents supplied in tablet form which enable the semiquantitative estimation of ketones in urine. It is based on Rothera's test. Acetest is manufactured by Ames.
See also: **Rothera's test**

Further reading: Kutter, D. (1977). *Rapid Clinical Diagnostic Tests.* (Munich–Vienna–Baltimore: Urban and Schwarzenburg).

ACETOACETATE

A compound which results from the breakdown of fatty acids. When carbohydrate metabolism is decreased, as in diabetes mellitus or starvation, the body can derive energy from triglyceride breakdown. The free fatty acids released from the breakdown of the triglycerides are metabolized to acetyl CoA, the excess acetyl CoA molecules condensing to form acetoacetate. Some acetoacetate is converted to acetone by spontaneous decarboxylation and the remainder is converted enzymically to β-hydroxybutyrate. Acetoacetate, acetone and β-hydroxybutyrate are collectively known as ketone bodies.

See also: **ketone bodies**

ACETONE

A ketone body formed by the spontaneous decarboxylation of acetoacetate.

See also: **ketone bodies**

ACETYLCHOLINESTERASE

See: **cholinesterase**

ACHLORHYDRIA

The absence of gastric hydrochloric acid secretion even after administration of stimulants such as pentagastrin or histamine. Achlorhydria can occur in gastritis, gastric carcinoma, pernicious anaemia and hypothyroidism.

ACID-BASE BALANCE

During the course of metabolism, hydrogen ions are continually being produced but these normally do not alter the pH of the blood because of a combination of several homeostatic mechanisms. These can be briefly summarized as:

(a) The blood buffering systems (although this is only a temporary measure).

(b) The elimination of CO_2 from the lungs.

(c) The renal excretion of hydrogen ions.

When these homeostatic mechanisms are impaired, acid-base disturbances result.

See also: **acidosis, alkalosis** (disturbances of acid–base balance), **bicarbonate, carbon dioxide,** (parameters used in the assessment of acid–base balance), **pH** (for the relationship between pH, bicarbonate and carbon dioxide)

Further reading: General list of clinical textbooks

α_1-ACID GLYCOPROTEIN (OROSOMUCOID, SEROMUCOID)

An α_1-globulin containing about 40% carbohydrate and having a molecular weight of 44 000. It is one of the acute phase proteins and increased blood levels are found in inflammatory conditions such as rheumatoid arthritis. Its exact role in the normal physiology of the body has not yet been elucidated, although it has been shown to bind certain drugs.

See also: **acute phase proteins**

Further reading: Editorial (1979). Drug binding to α_1-acid glycoprotein–Clinically important? *Lancet,* **1**, 368

ACIDOSIS

A condition in which there is loss of base or accumulation of acid capable of causing a fall in pH to below normal limits (an uncompensated acidosis). If this has been corrected by compensatory mechanisms (see below) it is known as a compensated acidosis. Acidosis can be classified into either a primary metabolic or respiratory disorder, although occasionally mixed types may occur.

Metabolic or non-respiratory acidosis

This can be caused by one or more of the following processes:

(1) Excessive production of acids, e.g. as in diabetic keto-acidosis.

(2) Reduced excretion of acids, e.g. renal failure.

(3) Increased loss of bicarbonate base, e.g. duodenal loss in diarrhoea.

Respiratory acidosis

This is caused by diseases in which there is decreased elimination of CO_2 through the lungs, e.g. bronchopneumonia, pulmonary emphysema.

In both types of acidosis, the pH can be restored to normal by the following compensatory mechanisms:

(1) The blood buffer systems.

(2) Increased respiration by stimulation of the respiratory centre, caused by the low pH. This results in more CO_2 being eliminated.

(3) Increased excretion of acids (providing renal function is normal).

See also: **acid–base balance, alkalosis, pH**

Further reading: Davenport, H.W. (1974). *The ABC of Acid–Base Chemistry*. 6th Edn., (Chicago: The University of Chicago Press)

Siggaard-Anderson, O. (1974). *The Acid–Base Status of the Blood,* 4th Edn. (Copenhagen: Munksgaard)

ACID PHOSPHATASES

A group of enzymes which hydrolyse phosphate esters at optimal pHs of below 7.0. Acid phosphatases are found in erythrocytes, platelets, liver, spleen, bone and, in males, in the prostate gland. Prostatic acid phosphatase constitutes about one third of the total acid phosphatase activity present in serum from normal male subjects.

Increases in serum prostatic acid phosphatase

(1) In prostatic carcinomas, especially with metastases.

(2) Slight increases can occur after rectal examination, passing a catheter, or in constipation.

Increases in serum non-prostatic acid phosphatase

(1) Carcinomatous bone deposits.

(2) Paget's disease of bone.

(3) Gaucher's and Niemann–Pick diseases.

(4) In some haematological disorders, e.g. myelocytic leukaemia.

(5) Hyperparathyroidism.

Measurement

Most of the substrates that have been used in measuring alkaline phosphatase activity have also been used to measure acid phosphatase, e.g. *p*-nitrophenylphosphate, phenylphosphate and α-naphthyl phosphate. In most cases of acid phosphatase estimation, it is the level of the prostatic phosphatase which needs to be known, and so specific inhibitors are included in the reaction mixtures. The prostatic isoenzyme is strongly inhibited by tartrate and so in many methods it is tartrate-labile acid phosphatase which is measured. On the other hand, the red cell enzyme which contributes significantly to the total serum activity is inhibited by formaldehyde and cupric ions. Many laboratories therefore measure formaldehyde-stable acid phosphatase as an indication of prostatic acid phosphatase.

Further reading: Bodansky, O. (1972). Acid phosphatase. In Bodansky, O. and Latner, A.L. (eds.) *Advances in Clinical Chemistry*. Vol. 15. p. 44. (New York: Academic Press Inc.)

ACROMEGALY

The condition which results from excess growth hormone secretion in adults, the usual cause being a pituitary adenoma. The condition is characterized by increases in bulk of the bones and soft tissues, especially of the jaw, hands and feet. Because growth hormone is involved in glucose homeostasis, impaired glucose tolerance can occur. Acromegaly can be diagnosed biochemically by:

(1) The finding of an abnormally high basal growth hormone level.

(2) The failure of growth hormone levels to fall during the course of a glucose tolerance test.

See also: **human growth hormone**

Further reading: Tunbridge, W.M.G. (1976). Acromegaly. *Br. J. Hosp. Med.*, **16**, 612

ACTH (ADRENOCORTICOTROPHIC HORMONE CORTICOTROPHIN)

A polypeptide hormone, secreted by the anterior pituitary, and comprised of 39 amino acids of which only the N-terminal 24 are needed for biological activity. ACTH stimulates glucocorticoid production by the adrenal cortex and also has a slight effect on adrenal androgen production, as well as a slight melanocyte stimulating activity. The secretion of ACTH is controlled by the hypothalamic releasing hormone, CRF (corticotrophin releasing factor). ACTH levels exhibit a circadian rhythm, being generally highest in the morning and lowest around midnight.

The N-terminal 24 amino acids which are essential for biological activity have been synthesized (Synacthen) and are used extensively in diagnosis and treatment.

Increases in serum levels

Increases are found in pituitary-dependent Cushing's syndrome, Nelson's syndrome, Addison's disease, congenital adrenal hyperplasia and ectopic ACTH production from a tumour.

Decreases in serum levels

Decreases are found in patients on steroid therapy, in panhypopituitarism and in cases of adrenal carcinoma.

Measurement

Although ACTH can be measured by bioassay (e.g. by cytochemical hormone assay) it is usually measured by radioimmunoassay.

Further reading: Hall, R., Anderson, J., Smart, G.A. and Besser, M. (1974). *Fundamentals of Clinical Endocrinology* 2nd Edn. (London: Pitman Medical)

Editorial (1977). ACTH secreting lung tumours. *Br. Med. J.*, **1**, 1047

ACTUAL BICARBONATE

A method of expressing whole blood bicarbonate. It can be calculated from the Henderson–Hasselbalch equation:

$$\mathrm{pH} \propto \frac{\text{bicarbonate concentration}}{P\mathrm{CO}_2}$$

if the pH and the $P\mathrm{CO}_2$ have been measured.

See also: **bicarbonate**

ACUTE INTERMITTENT PORPHYRIA (SWEDISH GENETIC PORPHYRIA, PYROLLOPORPHYRIA)

An inherited type of hepatic porphyria which is transmitted by a dominant gene. It is probably the most commonly occurring of the porphyrias in the UK. The disease is characterized by symptomless periods, punctuated by periods of abdominal pain which may be accompanied by neurological signs. These acute attacks can be precipitated by certain drugs especially barbiturates and sulphonamides. It can be diagnosed biochemically by the detection of increased amounts of δ-aminolaevulinic acid and porphobilinogen in the urine during acute attacks.

Further reading: Meyer, U.A. and Schmid, R. (1978). The porphyrias. In Stanbury, J.B. Wyngaarden, J.B., and Fredrickson, D.S. (eds.) *The Metabolic Basis of Inherited Disease.* 4th Edn., p. 1166. (New York: McGraw-Hill)

ACUTE PHASE PROTEINS

Proteins, mainly α_1 and α_2 globulins, which occur in elevated amounts in the blood in conditions in which there is active tissue damage (the acute phase reaction). Among these proteins are α_1-antitrypsin, α_1-acid glycoprotein and haptoglobin. γ-globulin levels may also be increased, and other changes include falls in albumin and prealbumin levels. A very sensitive indicator of the acute phase reaction is C-reactive protein, the levels of which can be increased several hundred fold in the serum.

Further reading: General list of clinical textbooks

ACUTE RENAL FAILURE

See: **renal failure**

ADDISON'S DISEASE

Primary adrenocortical hypofunction caused by destruction of the adrenal cortex, for instance by an autoimmune process or by tuberculosis.

Features

Destruction of the adrenal cortex results in a deficiency of glucocorticoids, mineralocorticoids and androgens, and the features of the disease are a reflection of this. Mineralocorticoid deficiency leads to sodium deficiency, which in turn leads to dehydration and this can present as an Addisonian crisis. The deficiency of glucocorticoids leads to hypoglycaemia and pronounced insulin sensitivity with a flat glucose tolerance curve. Because of the glucocorticoid deficiency, the pituitary produces maximal amounts of ACTH and MSH which may lead to skin pigmentation.

Diagnosis

Low plasma cortisol levels and high levels of ACTH are found. One of the essential findings is that the patient does not respond in the Synacthen stimulation test. Other biochemical findings are hypoglycaemia and hyponatraemia.

Further reading: General list of clinical textbooks.

ADENYLATE KINASE (MYOKINASE)

An enzyme which catalyses the reaction

$$\text{ATP} + \text{AMP} \rightleftharpoons 2\text{ADP}$$

Because it occurs mainly in skeletal muscle, raised serum levels can occur in muscle disease.

Further reading: Wilkinson, J.H. (1976). *The Principles and Practice of Diagnostic Enzymology*. (London: Edward Arnold)

ADRENAL GLAND

An endocrine gland situated above each kidney. Anatomically it can be divided into the medulla, which is part of the sympathetic nervous system, and the cortex. The cortex can be sub-divided into the outer layer, the zona glomerulosa which secretes aldosterone, and the zona fasciculata and zona reticularis which secrete most of the adrenocortical hormones. Adrenaline is secreted by the adrenal medulla.

ADRENALINE (EPINEPHRINE)

A hormone secreted by the adrenal medulla. It stimulates the sympathetic nervous system causing bronchodilatation, vasoconstriction and increased heart rate, which together raise the blood pressure. Adrenaline also increases glycogenolysis and lipolysis and this can increase blood levels of glucose and fatty acids.

Increased levels of adrenaline are secreted in tumours of the sympathetic nervous system (phaeochromocytoma and neuro-

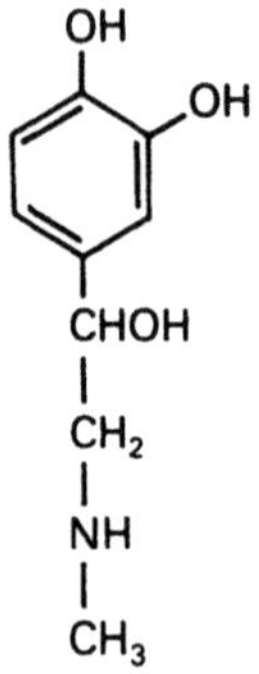

blastoma). It is usually measured in urine as its metabolite, 4-hydroxy-3-methoxymandelic acid (HMMA, VMA).

See also: **catecholamines**

ADRENOGENITAL SYNDROME

The syndrome associated with congenital adrenal hyperplasia. In this condition there is a deficiency of one of the enzymes involved in cortisol biosynthesis. ACTH levels are therefore maximal because of the lack of feedback control in the absence of cortisol. ACTH stimulates the overproduction of androgens. This can lead to sexual precosity in the male infant and masculinization in the female.

See also: **congenital adrenal hyperplasia**

ADSORPTION CHROMATOGRAPHY

A form of chromatography by which a mixture of components can be resolved by their differential ability to adsorb onto the surface of a solid stationary phase such as silica or alumina. The stationary phase may be packed in a column or coated on a thin layer plate. The mobile phase can either be liquid or gas. Although the apparatus used in adsorption chromatography can be the same as that used in partition chromatography, it should be remembered that there is an important theoretical difference between the two techniques.

Further reading: General list of analytical textbooks

AETIOCHOLANOLONE

One of the end products of androgen metabolism. It is excreted in the urine and can be measured as part of the 17-oxosteroid group of compounds.

See also: **17-oxosteroids**

AFFINITY CHROMATOGRAPHY

A form of chromatography in which molecules can be isolated according to their biological functions rather than on their physical and chemical properties which are the basis of other forms of chromatography. In affinity chromatography a binding molecule is covalently coupled to an insoluble matrix. The binding molecule is then able to absorb selectively from solutions the substance to be isolated. By changing the conditions, e.g. pH, the binding can be loosened and the substance can be eluted. Among the biologically specific reactions which can be utilized in this technique are antibody–antigen, and enzyme–inhibitor interactions.

Further reading: Cuatrecasas, P. and Anfinsen, C.B. (1971). Affinity chromatography. *Annu. Rev. Biochem.*, **40**, 259

AFIBRINOGENAEMIA

Congenital afibrinogenaemia is a rare hereditary disease in which there is little or no detectable fibrinogen in the blood. Patients with this condition have a bleeding tendency. It is inherited as an autosomal recessive.

Further reading: Ratnoff, O.D. (1978). Hereditary disorders of haemostasis. In Stanbury, J.B., Wyngaarden, J.B. and Fredrickson, D.S. (eds.) *The Metabolic Basis of Inherited Disease*, 4th Edn. p. 1755. (New York: McGraw-Hill)

AGAMMAGLOBULINAEMIA

A rare condition in which there is almost a complete absence of immunoglobulins in the serum, although cellular immunity appears to be normal. One syndrome, Bruton's disease, is a sex-linked recessive condition which affects male children with an incidence of 1 in 100 000 male births. An even rarer non-sex-linked agammaglobulinaemia has been described. Patients with agammaglobulinaemia suffer from repeated infections and require replacement therapy with human immunoglobulins.

Defective synthesis of only one or two of the major immunoglobulin classes can occur, the most frequent being IgA deficiency. These are known as dysgammaglobulinaemias.

Further reading: Rosen, F.S. and Merler, E. (1978). Genetic defects in gamma globulin synthesis. In Stanbury, J.B., Wyngaarden, J.B. and Fredrickson, D.S. (eds.) *The Metabolic Basis of Inherited Disease,* 4th Edn., p. 1726. (New York: McGraw-Hill)

AGAR

A polysaccharide obtained from certain seaweeds which forms a gel when water is added. It is composed of at least two fractions, agaropectin and agarose. Agar gels are used in many immunochemical precipitation reactions and can also be used as an electrophoretic medium, although agarose is preferred for this particular purpose.

See also: **agarose**

AGAROSE

A purified fraction of agar which has had agaropectin removed. Agaropectin contains sulphate and carboxylic acid groups which cause considerable electroendosmosis when agar gel is used as an electrophoretic medium. Agarose itself is neutral and therefore electroendosmosis is considerably reduced, making it an excellent electrophoretic medium.

See also: **agar**

AGGLUTINATION REACTIONS

Reactions which can be used for the detection of antigens located on the surface of cells (e.g. erythrocytes) or particles (e.g. latex particles). In the presence of antibody, agglutination occurs because of the cross-linking of the cells or particles by the antibody.

See also: **haemagglutination techniques, tanned red cell techniques**

β-ALANINAEMIA

An inborn error of metabolism in which there is a deficiency of the enzyme β-alanine transaminase. β-Alanine and γ-aminobutyric acid accumulate in the blood. Mental retardation is a feature of the condition.

Further reading: Scriver, C.R., Nutzenadel, W. and Perry, T.L. (1978). Disorders of β-alanine and carnosine metabolism. In Stanbury, J.B. Wyngaarden, J.B. and Fredrickson, D.S. (eds.) *The Metabolic Basis of Inherited Disease*, 4th Edn. p. 528. (New York: McGraw-Hill)

ALANINE AMINOTRANSFERASE (AlaAT, ALAT, ALT) (SERUM GLUTAMATE-PYRUVATE TRANSAMINASE, SGPT; GLUTAMATE-PYRUVATE TRANSAMINASE, GPT).

An enzyme found in high concentrations in the liver. Smaller amounts are also found in heart, kidney and skeletal muscle. It catalyses the reaction:

$$\begin{matrix}\text{alanine} \\ + \\ \text{2-oxoglutarate}\end{matrix} \rightleftharpoons \begin{matrix}\text{pyruvate} \\ + \\ \text{glutamate}\end{matrix}$$

Increases in the serum level of this enzyme are usually found in liver conditions especially hepatitis.

Measurement

(1) In the colorimetric, end-point, Reitman–Frankel method, the 2,4-dinitrophenylhydrazone derivative of pyruvate is measured. (The 2,4-dinitrophenylhydrazone derivative of 2-oxoglutarate is also present but this is not as chromogenic.) This method is now little used.

(2) Most laboratories measure the enzyme in a kinetic manner by a coupled enzyme reaction system. The pyruvate formed by alanine transaminase is converted to lactate by including lactate dehydrogenase in the reaction mixture. This is accompanied by the oxidation of NADH to NAD which can be followed spectrophotometrically at 340 nm.

See also: **DeRitis ratio**

Further reading: Wilkinson, J.H. (1976). *The Principles and Practice of Diagnostic Enzymology.* (London: Edward Arnold)

ALBINISM

A condition in which there is an absence of the pigment melanin. A congenital deficiency of the enzyme tyrosinase, which converts tyrosine to dihydroxyphenylalanine, an intermediate in melanin formation, causes one form of albinism. It has a recessive mode of inheritance.

Further reading: Witcop, C.J. Jr., Quevedo, W.C. Jr., and Fitzpatrick, T.B. (1978). Albinism. In Stanbury, J.B., Wyngaarden, J.B. and Fredrickson, D.S. (eds.) *The Metabolic Basis of Inherited Disease,* 4th Ed., p. 283. (New York: McGraw-Hill)

ALBUMIN

A protein having a molecular weight of approximately 65 000. It has a transport role in the blood since it can reversibly bind long chain fatty acids, bilirubin, calcium and certain hormones, e.g. thyroxine and cortisol. In addition to its transport functions it can serve as a reserve store of protein and contributes significantly to plasma colloidal osmotic pressure. Two congenital disorders of albumin synthesis have been described, analbuminaemia when there is deficient synthesis, and bisalbuminaemia when two types of albumin occur.

Increase in blood levels

Other than by haemoconcentration, high serum albumin levels are never found.

Decrease in blood levels

Decreased blood levels are found in three main types of pathological conditions:

(1) Decreased synthesis as in liver disease, malabsorption or malnutrition.

(2) Loss of protein from the body, e.g. nephrotic syndrome, protein losing enteropathy, and burns.

(3) Low serum albumin levels are found in a whole variety of

non-specific conditions (e.g. infections, inflammatory states) due to increased catabolism.

In addition to these pathological conditions, low serum albumin levels can be found as a result of recumbent posture, and also as a result of haemodilution e.g. in the later stages of pregnancy. Low levels are also found in neonates.

Measurement

(1) Dye binding methods. Certain dyes, e.g. bromocresol green (BCG) and 2-(4′-hydroxyazobenzene)-benzoic acid (HABA), bind to albumin, resulting in a change in the spectral characteristics of the dye, i.e. there is a shift in the wavelength of light at which maximum absorption occurs. Optical density measurements at the wavelength of maximum absorption of the bound dye can therefore be used to estimate the amount of albumin present.
Disadvantages of this method include interference by bilirubin and certain drugs which may displace the dye from the albumin binding sites resulting in falsely low values. At low serum albumin levels, results may be falsely high due to attachment of the dye to other serum proteins. One of the major advantages of dye binding methods is their ease of automation.

(2) Immunochemical methods using specific albumin antiserum. Radial immunodiffusion, electroimmunodiffusion and nephelometric techniques can all be used.

(3) Albumin can also be estimated densitometrically by electrophoresis of the serum followed by staining and scanning of the albumin band.

(4) Isolation of the albumin fraction by salt fractionation followed by protein determination by the biuret or Kjeldahl techniques.This is now a rarely used method.

(5) Albumin by difference. Determination of total protein and measurement of the globulin fraction by a method specific for globulins, e.g. the Hopkins–Cole reaction for tryptophan, enables the albumin to be estimated by difference.

Further reading: Peters, T. Jr., (1970). Serum albumin. In Bodansky, O. and Stewart, C.P. (eds.) *Advances in Clinical Chemistry*. Vol. 13, p. 37. (New York: Academic Press)
Slater, L., Carter, P.M. and Hobbs, J.R. (1975).

Measurement of albumin in the sera of patients. Technical Bulletin No. 34. *Ann. Clin. Biochem.*, **12**, 33

ALBUSTIX

A dipstick test manufactured by Ames for the estimation of protein in urine. It is based on the change of colour of tetrabromophenol blue from yellow to green caused by proteins, especially albumin. Alkaline urines may give false positive results.

Further reading: Kutter, D. (1977). *Rapid Clinical Diagnostic Tests*. (Munich–Vienna–Baltimore: Urban and Schwarzenberg)

ALCOHOL

See: **ethanol**

ALCOHOL TEST MEAL

A test which can be used to assess the ability of the stomach to secrete hydrochloric acid in response to a stimulant, in this case alcohol.

ALDOLASE

An enzyme found in heart, skeletal muscle and, to a lesser extent, the liver. It catalyses the reaction:

$$\text{fructose-1,6-diphosphate} \rightleftharpoons \text{dihydroxyacetone phosphate} + \text{glyceraldehyde-3-phosphate}$$

Increased serum levels occur in muscle diseases, acute hepatitis and following myocardial infarction.

Further reading: Wilkinson, J.H. (1976). *The Principles and Practice of Diagnostic Enzymology*. (London: Edward Arnold)

ALDOSTERONE

A mineralocorticoid secreted by the zona glomerulosa of the adrenal cortex. It is involved in sodium–potassium exchange across all cell membranes, including the distal renal tubules,

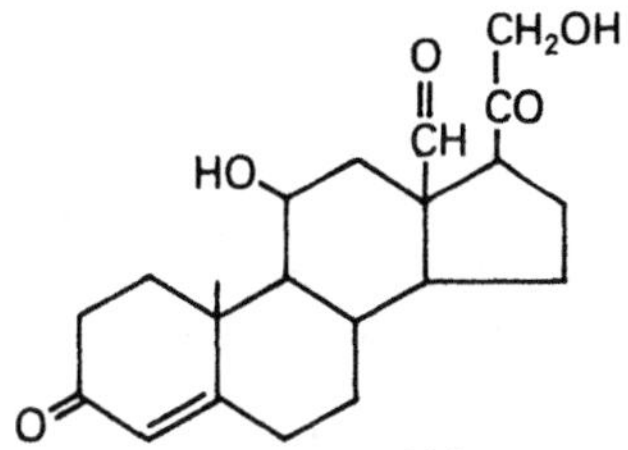

Aldosterone

where it increases sodium reabsorption in exchange for potassium. The renin–angiotensin system is a major factor in the control of aldosterone secretion. Aldosterone can be determined in either plasma or a 24-hour urine collection and is usually measured by radioimmunoassay.

See also: **aldosterone production rate, aldosteronism, renin–angiotensin sytem**

Further reading: General list of clinical textbooks

ALDOSTERONE PRODUCTION RATE

The determination of a patient's ability to synthesize aldosterone. Isotopically labelled aldosterone is given to the patient and the specific activity of an aldosterone metabolite is measured in a 24 hour urine. This test is used in situations where aldosterone measurements in urine may be misleading. e.g. when renal function is abnormal.

See also: **aldosterone**

ALDOSTERONISM

Primary (Conn's syndrome)

This is the secretion of excess aldosterone, due in most cases to a benign aldosterone-secreting adenoma of the adrenal cortex. Excess aldosterone causes sodium retention at the expense of potassium, and hypokalaemia develops.

Secondary

This occurs secondary to conditions in which there is reduced renal blood flow. These conditions include oedematous states (e.g. liver disease, nephrotic syndrome and protein malnutrition), damage to the renal vessels (e.g. renal artery abnormalities) and cardiac failure. The reduced renal blood flow stimulates the renin–angiotensin system which in turn stimulates aldosterone secretion. This results in sodium retention which stimulates ADH secretion. Consequently water retention develops. The biochemical findings in secondary aldosteronism therefore can include a low plasma sodium concentration and sometimes hypokalaemia.

See also: **aldosterone, potassium, renin–angiotensin system**

Further reading: General list of clinical textbooks

ALKALINE PHOSPHATASE

A group of enzymes which hydrolyse phosphate esters at optimal pHs of about 10. Alkaline phosphatase isoenzymes occur in many organs including intestine, bone (particularly osteoblasts), placenta, and liver and are also found in some patients with cancer (Regan isoenzyme).

Increases in serum levels

Increased serum alkaline phosphatase activity occurs mainly in hepatobiliary and bone conditions.

(1) *Hepatobiliary conditions*: Alkaline phosphatase is found in cells lining the bile canaliculi and is normally excreted in the bile. If the biliary flow is obstructed, the enzyme is regurgitated back into the bloodstream. Its presence in grossly elevated amounts in the serum therefore suggests cholestasis rather than liver cell damage. Large increases in serum alkaline phosphatase activity are therefore found in extra-hepatic biliary obstruction whereas more moderate increases are found in a whole variety of other liver conditions such as hepatitis, cirrhosis and hepatic carcinomas.

(2) *Bone conditions*: Elevated levels are found in those bone diseases which have increased osteoblastic activity, the highest levels being found in bone cancers and in Paget's disease (osteitis deformans). Moderate increases are found in osteomalacia, rickets and primary and secondary hyperparathyroidism. The raised serum alkaline phosphatase of normal children is due to increased osteoblastic activity.

(3) *Other conditions*: In addition to bone and liver conditions

alkaline phosphatase is also raised in other disorders such as Hodgkin's disease, ulcerative colitis and congestive heart failure. Elevated levels are also found in pregnancy.

Measurement

A multitude of methods exist for alkaline phosphatase measurement using a variety of substrates and conditions. This accounts for the variety of different units in which alkaline phosphatase activity can be expressed.

(1) The Bodansky method uses *β*-glycerophosphate as the substrate and measures the amount of free phosphate liberated.

(2) In the King–Armstrong method phenylphosphate is used as the substrate. The phenol liberated can be measured by Folin–Ciocalteu reagent, or in the case of the Kind and King modification, by 4-aminoantipyrene.

(3) In some methods, substrates are used which yield coloured products, e.g. *p*-nitrophenylphosphate (in the Bessey–Lowry–Brock method), thymolphthalein phosphate or phenolphthalein phosphate. These substrates can be used for the kinetic measurement of enzymic activity, unlike the previous methods which are end-point assays.

(4) *α*-Naphthol monophosphate can also be used in continuous monitoring procedures. The *α*-naphthol liberated absorbs light at 340 nm, the same wavelength at which NAD/NADH dehydrogenase reactions are measured.

Recently a number of phosphate accepting buffers have been used in alkaline phosphatase assays, onto which the liberated phosphate group is attached. These are amino alcohols such as diethanolamine, tris and ethylaminoethanol. When present they can enhance the rate of reaction.

Some divalent ions such as Mg^{2+} or Mn^{2+} activate the enzyme and are sometimes included in reaction mixures.

Isoenzyme determination

There have been two approaches to the routine determination of isoenzyme activity in serum. The first is the separation of the enzymes by electrophoretic means on a variety of media including starch, agar, polyacrylamide and cellulose acetate.

However the resolution between the liver and bone isoenzymes is poor.

The other approach is the selective inactivation of specific isoenzymes. Placental alkaline phosphatase, for instance, is remarkably stable to heat inactivation. Incubation of the enzyme at 65 °C has no effect on its activity, unlike the other isoenzymes which are inactivated. Other isoenzymes can be differentiated by their stability in other conditions. For instance phenylalanine inhibits placental and intestinal isoenzymes but has little effect on the bone and liver isoenzymes.

Further reading: Fishman, W.H. and Ghosh, N.K. (1967). Isoenzymes of human alkaline phosphatase. In Bodansky, O. and Stewart, C.P. (eds.) *Advances in Clinical Chemistry*, Vol. 10, p. 256. (New York: Academic Press).

Wilkinson, J.H. (1976). *The Principles and Practice of Diagnostic Enzymology*. (London: Edward Arnold)

ALKALINE TIDE

The secretion of hydrochloric acid by the stomach is accompanied by an increase in blood bicarbonate and pH.

ALKALI RESERVE

A method by which the level of plasma bicarbonate can be expressed. The plasma sample is equilibrated with 5% CO_2. This is an attempt to replace the CO_2 which may have escaped from the specimen during handling. (Hence its alternative name–CO_2 combining power). The sample is treated with acid and the total CO_2 liberated gives an indication of the plasma bicarbonate level.

See also: **bicarbonate**

ALKALOSIS

A condition in which there is an accumulation of base or loss of acid capable of causing a rise in pH to above normal limits (an uncompensated alkalosis). If this has been corrected by compensatory mechanisms (see below), it is known as a compensated alkalosis. There are two classes of alkalosis depending on their aetiology, although occasionally a mixed type may be encountered.

Metabolic or non-respiratory alkalosis

This may be caused by one of the following processes:

(1) Excessive intake of alkali e.g. sodium bicarbonate.

(2) Excessive loss of hydrochloric acid from the stomach e.g. vomiting.

(3) In potassium depletion (e.g. Cushing's syndrome, aldosteronism). Potassium depletion in the extracellular fluid causes K^+ to move out of the cells, and this is accompanied by a movement of H^+ and Na^+ into the cells. This results in an extracellular alkalosis (although there is at the same time an intracellular acidosis).

Respiratory alkalosis

This can result from diseases in which there is increased elimination of CO_2 from the lungs as a result of increased respiration, e.g. hysteria, hypoxia or salicylate poisoning.

In both metabolic and respiratory alkaloses, the pH can be restored to normal by the following processes:

(1) Blood buffer systems.

(2) Depression of the respiratory centre of the brain by the high pH, resulting in less CO_2 being eliminated.

(3) Decreased renal Na^+ / H^+ exchange and decreased renal reabsorption of bicarbonate.

See also: **acid-base balance, acidosis, pH**

Further reading: Davenport, H.W. (1974). *The ABC of Acid–Base Chemistry*. 6th Edn. (Chicago: The University of Chicago)

Siggaard-Anderson, O. (1974). *The Acid–Base Status of the Blood*. 4th Edn. (Copenhagen: Munksgaard)

ALKAPTONURIA

A rare inborn error of metabolism in which there is a deficiency of the enzyme homogentisic acid oxidase. It is inherited as an autosomal recessive.

The oxidative breakdown of the amino acid tyrosine contains a stage in which homogentisic acid is converted to maleylacetoacetic acid, a reaction catalysed by homogentisic acid oxidase. A deficiency of this enzyme results in the accumulation of homogentisic acid in blood, tissues and urine. Oxidation and polymerization of the excessive amounts of homogentisic acid results in the formation of the black pigment alkapton which can be deposited in cartilages, causing darkening. This condition is called ochronosis and may eventually lead to arthritis. Urines from affected individuals darken on standing, due to the conversion of homogentisic acid to alkapton.

See also: **homogentisic acid**

Further reading: La Du, B.N. (1978). Alkaptonuria. In Stanbury, J.B., Wyngaarden, J.B. and Fredrickson, D.S. (eds.) *The Metabolic Basis of Inherited Disease,* 4th Edn., p. 268. (New York: McGraw-Hill)

ALLEN CORRECTION

A method of correcting optical density measurements when there is interference by background absorption. Optical density measurements are made at the wavelength of maximum absorption of the compound being examined and at two other wavelengths equidistant from the peak. The absorbance of the two latter wavelengths are averaged and subtracted from the peak reading to give the corrected reading. This is shown in the diagram.

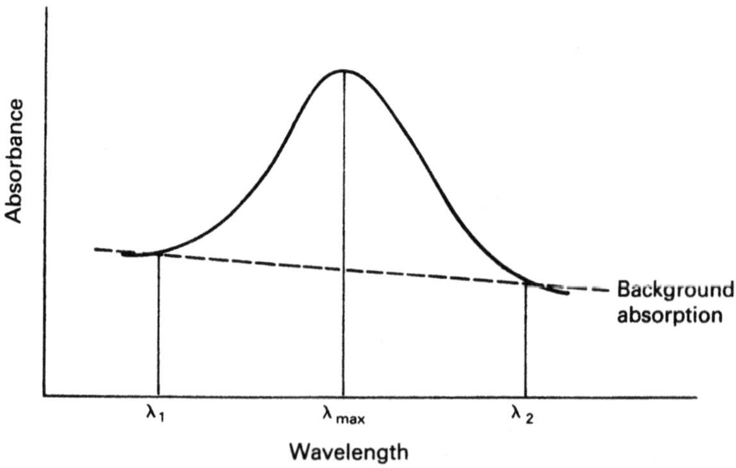

Corrected absorption at $\lambda_{max} = A_{max} - \left(\frac{A_1 + A_2}{2}\right)$

where A_{max}, A_1, A_2 = absorbances at the 3 wavelengths respectively

Such corrections are used in the determination of steroids by the Zimmermann reaction and in the spectrophotometric determination of porphyrins, salicylates and other compounds. Care must be taken whenever it is applied as it is possible that even larger errors might be introduced than there would be without the correction, e.g. when the absorbance of the background is not linear in the region measured.

Further reading: General list of analytical textbooks

ALLOPURINOL (HYDROXYPYRAZOLOPYRIMIDINE)

A drug used in the treatment of gout. Because of its structural similarity to hypoxanthine, it acts as a competitive inhibitor of the enzyme xanthine oxidase, thereby reducing uric acid production.

See also: **gout**

ALUMINIUM

It is possible that aluminium may be involved in the development of encephalopathy and also of osteomalacia in haemodialysed patients (aluminium salts are present in the dialysate water and aluminium hydroxide is administered to control the hyperphosphataemia). Aluminium is thought to accumulate in the bone and interfere with mineralization processes.

AMINO ACID ANALYSER

An instrument for the measurement of individual amino acids in an amino acid mixture. The amino acids are separated on an ion-exchange column and estimated colorimetrically on a continuous flow analyser after their reaction with ninhydrin. In some instruments the amino acids are measured fluorimetrically.

AMINOACIDURIA

An excess of amino acids in the urine. It may arise from an 'overflow' effect when there are raised blood levels of amino acids (as in phenylketonuria), or as a result of a renal disorder when the amino acids are not reabsorbed by the renal tubules.

Aminoacidurias may be generalized, when there is excessive secretion of several non-related amino acids, or specific, when there is excess excretion of a single amino acid or a related group of them (as in cystinuria or Hartnup disease).

p-AMINOHIPPURIC ACID (PAH)

A substance used in the assessment of renal function. At low plasma levels, estimation of its clearance can be used to measure renal plasma flow whereas at higher plasma levels it can give an indication of the tubular secretory capacity.

Further reading: Mitchell, F.L. Veall, N. and Watts, R.W.E. (1972). Scientific Review No. 2. Renal function tests suitable for clinical practice. *Ann. Clin. Biochem.*, **9**, 1

δ-AMINOLAEVULINIC ACID (δ-ALA)

An intermediate in the synthesis of porphyrins. It is formed by the condensation of glycine and succinate, a reaction catalysed by the enzyme δ-ALA synthetase. δ-ALA excretion in urine is increased in some types of porphyria and in lead poisoning. δ-ALA can be measured in urine and serum by its reaction with Ehrlich's reagent.

See also: **lead and lead poisoning**

δ-AMINOLAEVULINIC ACID DEHYDRATASE

The enzyme which converts δ-aminolaevulinic acid to porphobilinogen. Decreased red cell δ-aminolaevulinic acid dehydratase activity is found in cases of lead poisoning.

See also: **lead and lead poisoning**

AMITRIPTYLINE

A tricyclic antidepressant. Several simple colour reactions can be used for its detection in urine.

AMMONIA

This occurs in the blood as a result of (a) the deamination of amino acids and (b) the absorption of ammonia arising from the action of urea-splitting bacteria in the gut. It is a toxic substance and is therefore rapidly metabolized by liver enzymes to urea which is excreted in the urine.

Ammonia itself is involved in urinary buffering mechanisms. The action of glutaminase on glutamine in the distal tubular cells results in ammonia being secreted into the urine, where it buffers hydrogen ions, causing more acid to be secreted.

Increases in blood levels

High blood levels are found in advanced liver disease.

Measurement

(1) Acid–base titration.

(2) The Berthelot reaction.

(3) Using Nessler's reagent.

(4) By use of the enzyme glutamate dehydrogenase which converts 2-oxoglutarate and ammonia to glutamate, with a concomitant oxidation of NADH to NAD which can be followed spectrophotometrically.

$$\text{2-oxoglutarate} + NH_4 + NAD \longrightarrow \text{glutamate} + H_2O + NAD$$

Further reading: General list of clinical and analytical textbooks

AMMONIUM CHLORIDE LOADING TEST

A test which measures the ability of the kidney to secrete an acid urine. The ammonium ion (NH_4^+) behaves as an acid because it dissociates to ammonia and H^+. When ammonium chloride is ingested, the kidneys should normally secrete the excess hydrogen ions in the urine. Urinary pH, titratable acidity and ammonia content can be measured as an indication of the acid output. In normal subjects, the pH falls to 5.2 or below, whereas in renal tubular acidosis, no acidification occurs.

Further reading: Mitchell, F.L., Veall, N. and Watts, R.W.E. (1972). Scientific Review No. 2. Renal function tests suitable for clinical practice. *Ann. Clin. Biochem.*, **9**, 1

AMNIOTIC FLUID

The fluid which surrounds the fetus. In early pregnancy its composition resembles serum, while in the later stages it has a closer resemblance to fetal urine.

AMPEROMETRY

An analytical technique which is based on the measurement of current flowing through an electrical cell when a constant potential is applied to the electrodes. The Clark PO_2 electrode is an example of this technique.

See also: **oxygen**

AMPHETAMINES

A class of drugs, taken for their stimulant action on the brain. Continual dosage may result in addiction. They can be estimated in urine by gas–liquid chromatography, ultraviolet spectrophotometry or by their reaction with metanil yellow to form a coloured complex.

Further reading: General list of analytical textbooks

AMPHOLYTE

A molecule which can be either positively or negatively charged. These type of molecules are used in isoelectric focusing.

See also: **isoelectric focusing**

AMYLASE

An enzyme which catalyses the hydrolysis of high molecular weight carbohydrates such as starch and glycogen. Human amylases are referred to as α-amylases because they attack the α-1,4-hemiacetal linkages of starch in a random manner, as distinct from the β-amylases of plant and bacterial origin which can act only at the terminal reducing end of a carbohydrate chain. Hydrolysis of straight chain carbohydrates results in the formation of the disaccharide, maltose. In the case of the branched-chain carbohydrates such as amylopectin or glycogen, hydrolysis results in the formation of maltose and limit dextrin, since the enzyme does not attack the α-1,6-linkages at the branch points of these carbohydrates.

The enzyme present in normal serum is derived mainly from the pancreas and the salivary glands.

Increases in serum levels

(1) Pancreatic disease especially acute and chronic pancreatitis.

(2) Salivary gland disease, e.g. mumps.

(3) Macroamylasaemia, a symptomless condition in which complexes are formed between amylase and other plasma proteins, especially IgA. These high molecular weight complexes are not filtered at the glomerulus and accumulate in the plasma.

Decreases in serum levels

Lowered serum amylase levels are found in a variety of conditions but are of little clinical significance.

Measurement of amylase activity

(1) In the *saccharogenic* type of assay, enzyme activity is measured by the amount of reducing sugar formed. Any of the traditional substances for measuring reducing substances can be used, e.g. ferricyanide.
Amylase can be measured by kinetic methods using coupled enzyme assay systems. The maltose formed by the action of amylase is converted to glucose, by including maltase in the reaction mixture. Glucose oxidase which is also added converts glucose to gluconic acid as follows:

$$\text{Glucose} + O_2 \longrightarrow \text{Gluconic Acid} + H_2O_2$$

The overall reaction can be followed by measuring the amount of oxygen consumed in the reaction by a PO_2 electrode, or by using the H_2O_2 formed to oxidize an oxygen-accepting dye.

(2) In the *amyloclastic* type of assay, enzyme activity is followed by measuring the decrease in the substrate, starch.

Starch is measured by its ability to form a blue colour with iodine. Some methods involve measuring the amount of starch hydrolysed in a given time, while in other methods the amount of time needed to hydrolyse a given amount of starch is determined.

(3) A number of dye-labelled amylase substrates have been introduced in recent years by several companies. A coloured dye is coupled to starch substrate. Hydrolysis of the substrate by the enzyme results in the formation of water-soluble dye fragments which can be separated from the unhydrolysed substrate by centrifugation or filtration.

Units

Amylase activity can be expressed in a variety of units, e.g. Somogyi, Wohlgemuth or International Units, depending upon the method of assay.

Further reading: Wilkinson, J.H. (1976.) *The Principles and Practice of Diagnostic Enzymology*. (London: Edward Arnold)

AMYLOBARBITONE

An intermediate acting barbiturate.

See also: **barbiturates**

AMYLOID

An abnormal proteinaceous material composed in some cases of light chain residues from γ-globulins. It can be deposited in many of the body's organs especially the liver and kidneys. It is referred to as amyloid because of its starch-like reactions. Amyloid disease has been classified into two categories based on the composition of the protein. The familial variety or the type associated with myeloma is referred to as primary amyloid. Secondary amyloid occurs in diseases other than myeloma e.g. rheumatoid arthritis, tuberculosis. Amyloidosis can be diagnosed by the Congo Red test, although this is potentially dangerous because of possible allergic reactions. In this test the dye is injected intravenously. Any amyloid material present binds the dye and its removal from the circulation can be followed by taking a number of blood samples up to an hour after injection.

Further reading: Glenner, G.G., Ignaczak, T.F. and Page, D.L. (1978). The inherited systemic amyloidoses and localized amyloid deposits. In Stanbury, J.B., Wyngaarden, J.B. and Fredrickson, D.S. (eds.) *The Metabolic Basis of Inherited Disease,* 4th Edn., p. 1308. (New York: McGraw-Hill)

Editorial (1979). Pathogenesis of amyloid disease. *Br. Med. J.,* **1**, 216

ANALBUMINAEMIA

A rare hereditary disease in which there is almost a complete absence of albumin in the serum. The condition is inherited as an autosomal recessive. There appear to be few pathological consequences of the albumin deficiency. Affected individuals often have raised serum globulin levels which may be a compensatory measure to maintain the colloid osmotic pressure.

Further reading: Bearn, A.G. and Litwin, S.D. (1978). Deficiencies of circulating enzymes and plasma proteins. In Stanbury, J.B., Wyngaarden, J.B. and Fredrickson, D.S. (eds.) *The Metabolic Basis of Inherited Disease* 4th Edn., p. 1712 (New York: McGraw-Hill)

ANDROGENS

These are a group of C-19 steroids which influence the male secondary sex characteristics. They are made by the testes, adrenals and to a lesser extent by the ovaries. The testes secrete testosterone while the main adrenal androgens are dehydroepiandrosterone and androstenedione. Urinary 17-oxosteroid estimations give an indication of androgen production.

See also: **17-oxosteroids, testosterone**

ANDROSTANEDIOL

An androgen produced in peripheral tissue from testosterone.

ANDROSTENEDIONE

An intermediate in the synthesis of testosterone. Together with dehydroepiandrosterone it is one of the principal androgens secreted by the adrenal gland.

ANDROSTERONE

One of the end products of androgen metabolism. It is one of the compounds measured in the 17-oxosteroid determination.

See also: **17-oxosteroids**

ANGIOTENSIN I

A decapeptide formed by the action of renin on angiotensinogen. It is converted to the octopeptide angiotensin II by a peptidase located mainly in the lungs.

See also: **aldosterone, renin–angiotensin sytem**

ANGIOTENSIN II

An octopeptide hormone which is produced as a result of the action of a peptidase on angiotensin I. Its two main actions are the stimulation of aldosterone production from the adrenal cortex, and an action on the blood vessel walls causing vasoconstriction. It can be measured by radioimmunoassay.

See also: **aldosterone, renin–angiotensin system**

ANGIOTENSINOGEN

A plasma protein which is converted to angiotensin I by the action of renin.

See also: **aldosterone, renin–angiotensin sytem**

ANHAPTOGLOBINAEMIA

A congenital condition in which there is an absence of haptoglobin in the blood. It is a relatively common inborn error, being found in approximately 4% of negroes.

Further reading: Bearn, A.G. and Litwin, S.D. (1978). Deficiencies of circulating enzymes and plasma proteins. In Stanbury, J.B., Wyngaarden, J.B. and Fredrickson, D.S. (eds.) *The Metabolic Basis of Inherited Disease,* 4th Edn., p. 1712. (New York: McGraw-Hill)

ANION GAP

The plasma electrolytes most commonly measured are sodium, potassium, chloride and bicarbonate. The sum of the chloride and bicarbonate is always less than the sum of the sodium and potassium. This is the anion gap and it is due to a number of other ions, e.g. phosphate, sulphate and organic acids such as lactate. Large anion gaps are found in renal failure and diabetic ketoacidosis.

Further reading: Editorial (1977). The anion gap. *Lancet*, **1**,785

ANODIC STRIPPING VOLTAMETRY

A technique which can be used for trace element analysis. It consists of depositing metals from the solution being investigated onto a mercury cathode. The mercury electrode is then made the anode and by altering the potential, the metals which have been deposited in the previous step are stripped off. The stripping is carried out under polarographic conditions and as each metal is removed in turn from the mercury, a peak appears on the current vs. voltage curve. The quantity of each metal is determined from the height of the peak.

Further reading: Ellis, W.D. (1973). *J. Chem. Educ.*, **50**, A131

ANTIBODY

An immunoglobulin made by animals in response to an antigen. The antibody combines specifically with the antigen and this plays an important role in the body's defence against infection or foreign substances. The specificity of the antibody–antigen reaction enables highly specific and sensitive assays to be performed for the estimation of many substances.

See also: **immunochemical techniques**

ANTICOAGULANT

A substance which prevents blood from clotting. They are used therapeutically in treating patients with a history of thrombosis and are used extensively in the laboratory to prevent blood specimens from coagulating when whole blood or plasma is required for analysis. The anticoagulants EDTA, potassium oxalate and sodium citrate act by chelating calcium which is required for blood clotting. Another anticoagulant, heparin, acts by antagonizing thrombin.

ANTIDIURETIC HORMONE (VASOPRESSIN, PITRESSIN)

A polypeptide hormone secreted by the posterior pituitary gland. It is involved in fluid homeostasis, exerting its action in the distal renal tubular cells where it increases passive water reabsorption along the osmotic gradient produced by the countercurrent mechanism. This results in a concentrated urine being produced. Deficient production of the hormone, or failure of the renal tubules to respond to it, results in diabetes insipidus. Alternatively inappropriate ADH secretion can occur in conditions such as infections or as a result of ectopic ADH secretion by a tumour. This results in water retention which reveals itself as hyponatraemia. ADH can be measured by radioimmunoassay.

See also: **countercurrent mechanism, diabetes insipidus**

Further reading: Scheiner, E. (1975). The relationship of antidiuretic hormone to the control of volume and tonicity in the human. In Bodansky, O., and Latner, A.L. (eds.) *Advances in Clinical Chemistry*, Vol. 17, p. 2. (New York: Academic Press)

ANTI-DNA ANTIBODIES

Circulating antibodies to DNA are found in connective tissue disorders such as systemic lupus erythematosus (SLE) and rheumatoid arthritis. Antibodies against double-stranded native DNA are found almost exclusively in the SLE patients whereas antibodies against single-stranded denatured DNA are found in other connective tissue disorders, in addition to SLE.

See also: **antinuclear factors**

Further reading: Holborow, E.J. (1978). The serology of connective tissue disorders. *Br. J. Hosp. Med.*, **19**, 250

ANTIGEN

A substance which is capable of stimulating the production of antibodies. The term is often used synonymously with immunogen, although strictly speaking the term antigen should refer to any substance which is capable of reacting with antibody (e.g. *in vitro*) without necessarily having the ability to stimulate antibody production *in vivo*.

ANTINUCLEAR FACTORS

These are antinuclear antibodies found in the serum in a number of autoimmune disorders, such as systemic lupus erythematosus, rheumatoid arthritis and chronic active hepatitis. These antibodies are active against different components of the cell nucleus.

See also: **anti-DNA antibodies**

Further reading: Holborow, E.J. (1978). The serology of connective tissue disorders. *Br. J. Hosp. Med.,* **19**, 250

ANTISERUM

Serum which contains antibodies. Antisera used in diagnostic tests are obtained from animals injected with the appropriate antigen. Antiserum reacting against several antigens is termed polyvalent. If it reacts against only one antigen it is termed monovalent (or monospecific)

α_1-ANTITRYPSIN

An $\alpha_1$globulin with a molecular weight of approximately 45 000. It inhibits proteolytic enzymes in the blood and it is one of the acute phase proteins. There are several genetic variants of the protein. A congenital deficiency of α_1-antitrypsin can occur producing the symptoms of infantile liver cirrhosis and early onset emphysema. The latter may be due to release of proteolytic enzymes when lung tissue is infected.

Measurement

(1) Immunochemical methods, e.g. radial immunodiffusion or electroimmunoassay.

(2) Enzymically by measuring trypsin activity with and without the addition of the patient's serum.

Further reading: Bearn, A.G. and Litwin, S.D. (1978). Deficiencies of circulating enzymes and plasma proteins. In Stanbury, J.B., Wyngaarden, J.B. and Fredrickson, D.S. *The Metabolic Basis of Inherited Disease,* 4th Ed., p. 1712 (New York: McGraw-Hill)

Editorial. (1977). Childhood liver disease with α_1-antitrypsin deficiency. *Lancet,* **1**, 82

APOFERRITIN

A protein which binds iron to form the iron storage compound ferritin.

See also: **ferritin**

ARGENTAFFIN CELLS

Cells which stain with silver salts and which are found in several sites in the alimentary tract. They synthesize the hormone serotonin.

See also: **carcinoid syndrome**

ARGENTAFFINOMA

A tumour of argentaffin cells which cause the condition known as carcinoid syndrome.

See also: **carcinoid syndrome**

ARGININE

A basic amino acid found in many proteins. It is also involved in urea biosynthesis. In the inborn error, cystinuria, it is excreted in the urine in large amounts, along with other basic amino acids.

See also: **cystinuria**

ARGININE STIMULATION TEST

A test which assesses the ability of the pituitary to secrete growth hormone. In normal subjects, growth hormone is secreted in response to amino acids. In this test arginine is infused intravenously and growth hormone levels are measured at half-hour intervals for 2 hours. Failure of growth hormone levels to rise significantly is suggestive of growth hormone deficiency.

See also: **human growth hormone**

ARGININOSUCCINICACIDURIA

An inborn error of metabolism which is due to a deficiency of the urea cycle enzyme argininosuccinase. Argininosuccinic acid and ammonia accumulate and the symptoms include mental retardation and hepatomegaly.

Further reading: Shih, V.E. (1978). Urea cycle disorders and other congenital hyperammonaemic syndromes. In Stanbury, J.B., Wyngaarden, J.B. and Fredrickson, D.S. (eds.) *The Metabolic Basis of Inherited Disease*, 4th Edn., p. 362. (New York: McGraw-Hill)

ARIBOFLAVINOSIS

The condition caused by deficiency of the vitamin riboflavin. Symptoms include rough scaly skin and oral, anal and vaginal lesions.

See also: **riboflavin**

ARSENIC

Poisoning by arsenic compounds is occasionally encountered in the laboratory. The toxic action of arsenic is due to its inhibition of sulphydryl enzymes. It can be detected in body fluids by the Reinsch test which is based on the reduction of arsenic to its elemental form by metallic copper in the presence of acid. The arsenic is deposited on the copper as a dark film.

Further reading: General list of analytical textbooks

ASCORBIC ACID (VITAMIN C)

A water soluble vitamin which cannot be synthesized by man and therefore has to be obtained from the diet. It is found extensively in vegetables and fruit, especially the citrus varieties. Since the vitamin is carried mainly in the leukocytes, its measurement in these cells gives some indication of the vitamin C status of the body. The ascorbic acid saturation test can also be used to assess the vitamin status. The biochemical role of the vitamin is obscure although it does seem to be required for collagen formation. Deficiency of the vitamin causes scurvy, the symptoms of which can be related to poor collagen formation. These include poor wound-healing, osteoporosis (due to bone matrix deficiency), a tendency to bleed (due to deficiences in the vascular walls) and anaemia.

Measurement

Ascorbic acid is usually estimated by its ability to reduce the dye 2,6-dichlorophenolindophenol to a colourless form.

Further reading: General list of clinical and analytical textbooks

ASCORBIC ACID SATURATION TEST

A test for assessing the vitamin C status of the body. Ascorbic acid is given orally and its output in the urine is measured. In normal subjects, more of the vitamin will be excreted in the urine, because the tissue stores are already saturated. In deficient patients, less vitamin will be excreted because of its uptake by the tissues.

See also: **ascorbic acid**

ASPARTATE AMINOTRANSFERASE (AspAT, ASAT, AST) (SERUM GLUTAMATE-OXALOACETATE TRANSAMINASE, SGOT, GLUTAMATE-OXALOACETATE TRANSAMINASE, GOT)

An enzyme, high levels of which are found in heart, liver, red blood cells and skeletal muscle. It catalyses the reaction:

$$\text{aspartate} + \text{2-oxoglutarate} \rightleftharpoons \text{oxaloacetate} + \text{glutamate}$$

Increases in serum levels

Increases are found in many liver diseases where there is liver cell destruction, especially hepatitis. It is also raised after myocardial infarction and in muscle diseases.

Measurement

(1) In the colorimetric, end-point, Reitman–Frankel method, oxaloacetate is measured as the 2,4-dinitrophenylhydrazone derivative. (The 2,4-dinitrophenylhydrazone derivative of 2-oxoglutarate is also formed but this is not as chromogenic.) This method is now little used.

(2) The enzyme can be measured in a kinetic manner by a coupled enzyme reaction system. The oxaloacetate formed by aspartate aminotransferase is converted to malate by including malate dehydrogenase in the assay system. This is accompanied by the oxidation of NADH to NAD which can be followed spectrophotometrically at 340 nm.

(3) Colorimetric methods can be used which depend upon the

reaction of oxaloacetate with a diazo dye, e.g. azoene fast violet B.

Further reading: Wilkinson, J.H. (1976). *The Principles and Practice of Diagnostic Enzymology*. (London: Edward Arnold)

ASPIRIN

See: **salicylate**

ASTRUP

A name which is applied to a technique of measuring blood acid–base parameters, devised by the Danish biochemist Poul Astrup. The name is also used to describe products of the Radiometer Corporation which measure blood acid–base parameters by the Astrup technique.

The basis of the technique is the inverse relationship between blood pH and PCO_2.

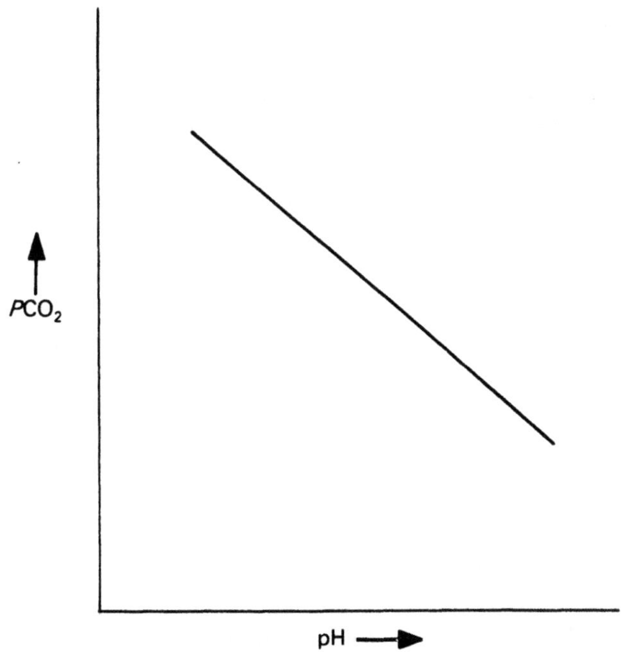

The pH of a sample of whole blood is first measured. The remainder is then equilibrated at two different concentrations of CO_2 (i.e. two different PCO_2s) and two more pH readings are made. A line is drawn between these last two pH readings and by placing the sample's original pH on the line, the patient's PCO_2 can be read off.

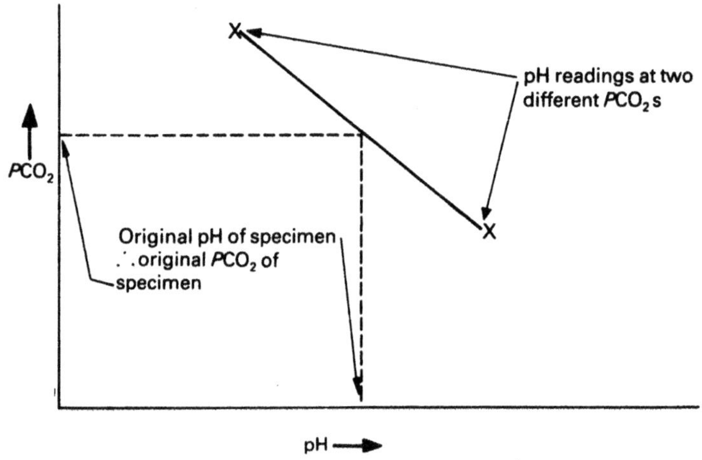

The standard bicarbonate is found at the point where the line intersects another line corresponding to a PCO_2 of 5.3 kPa (40 mmHg). Similarly the base excess is found at the point where the line intersects the base excess curve.

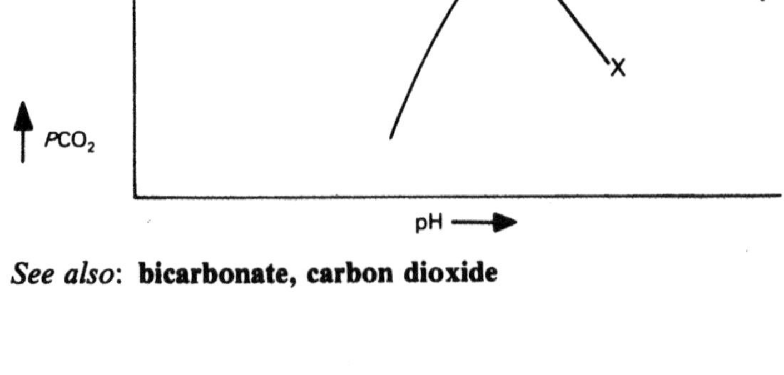

See also: **bicarbonate, carbon dioxide**

ATOMIC ABSORPTION

An analytical technique used for the measurement of many metallic elements, e.g. calcium, magnesium, lead, copper and iron.

Atomic absorption flame photometry

In this technique, a liquid sample is nebulized in a cloud chamber and passed into a flame, where the element is dissociated by the heat from its chemical bonds and placed in an unexcited or ground state. Narrow wavelength light from a hollow cathode lamp is passed through the flame and some of the ground state atoms are excited by the radiation. This results in a net decrease in the intensity of the beam and this can be measured by a photoelectric detector. The process is therefore analagous to absorption spectrophotometry for the measurement of molecules.

The hollow cathode lamp takes the form of a cup-shaped cathode in a space filled with an inert gas. The cathode is made of the same metal as the one which is to be estimated- e.g. when measuring calcium, a cathode made of metallic calcium is used. An electrical potential is applied between the anode and the cathode, resulting in the cathode being bombarded with gaseous ions which exchange energy with the element. The element emits radiation in discrete lines, but some of the secondary lines may be removed with filters. Double beam instruments are available which correct for variations in the output of the lamp.

Flameless atomic absorption

This technique can be used for assaying small volumes of material. The flame described above is replaced by a carbon rod or tantalum strip onto which the sample is placed. The temperature of the rod is raised electrically which in turn dries, ashes and atomizes the sample into the light path.

Further reading: Broughton, P.M.G. and Dawson, J.B. (1972). Instrumentation in clinical chemistry. In Bodansky, O. and Latner, A.L. (eds.) *Advances in Clinical Chemistry*. Vol. 15, p. 288. (New York: Academic Press)

ATOMIC EMISSION FLAME PHOTOMETRY

A technique used in the measurement of certain metal ions, particularly sodium, potassium, lithium and calcium. It is based on the emission of light of a characteristic wavelength by metal atoms when they are excited by the thermal energy of a flame.

When salts are introduced into a flame, they dissociate to give neutral atoms. Energy from the flame causes a small proportion of these to move into a higher energy state. When these excited atoms fall back into the ground state, light is emitted, the intensity of which is proportional to the number of atoms excited.

A flame emission photometer consists of an atomizer by which the sample is transformed into an aerosol spray before it is introduced into the flame. In the flame the metal atoms emit light of a wavelength characteristic to that element. The light passes through filters or diffraction gratings which isolate a single spectral line, and its intensity is measured by a photoelectric device. Internal standards can be used to compensate for variations in the intensity of the flame, e.g. lithium if sodium and potassium are being measured.

Further reading: Broughton, P.M.G. and Dawson, J.B. (1972). Instrumentation in clinical chemistry. In Bodansky, O. and Latner, A.L. (eds.) *Advances in Clinical Chemistry*. Vol. 15, p. 288. (New York: Academic Press)

AUSTRALIA ANTIGEN (HEPATITIS B SURFACE ANTIGEN, HBsAg)

An antigen present in the blood of some patients with viral hepatitis. It is a marker of the hepatitis B virus. Blood donors are routinely screened for the presence of Australia antigen, and if it is found, they are excluded from giving blood.

AUTOANALYZER

A name which is the registered trademark of Technicon Instruments and by which the continuous flow instruments of that company are known.

See also: **continuous flow analysis**

AUTOANTIBODY

An antibody produced by the body against one of its own components.

AUTOIMMUNE DISEASES

Diseases in which the body produces antibodies against its own components. Among the diseases which are thought to have an autoimmune basis are: primary myxoedema, thyrotoxicosis, pernicious anaemia, Addison's disease, Goodpasture's syndrome, myasthenia gravis, some haemolytic anaemias, primary biliary cirrhosis, active chronic hepatitis, ulcerative colitis, Sjøgren's syndrome, rheumatoid arthritis and systemic lupus erythematosus. See separate entries for some of these diseases.

AUTOMATED IMMUNE PRECIPITATION (AIP)

An automated technique for immunochemical assays, using continuous-flow instruments. The basis of the technique is the reaction of antigens in the patient's sera (e.g. immunoglobulins, transferrin, or other plasma proteins) with specific antisera. The immune complexes formed by the antigen–antibody reaction scatter light and this can be measured by passing the stream through a fluoronephelometer.

Further reading: Ritchie, R.F., Alper, C.A., Graves, J., Pearson, N. and Larson, C. (1973). Automated quantitation of proteins in serum and other biologic fluids. *Am. J. Clin. Pathol.*, **59**, 151

White, P.A.E. and Strong, R. (1979). Automated immunoprecipitation and laser nephelometry, In Milford Ward, A. and Whicher, J.T. (eds.) *Immunochemistry in Clinical and Laboratory Medicine* p. 23 (Lancaster: MTP Press)

AUTOMATION

The performance of laboratory tests by machines which are capable of carrying out some or all of the procedures which would otherwise have to be carried out manually. Such procedures include pipetting, centrifugation, measurement of optical density (or some other parameter) and calculations.

See also: **centrifugal analysers, continuous flow analysis, discrete analysis**

AZOSTIX

A reagent stick manufactured by Ames for the estimation of blood urea. The basis of the test is the hydrolysis of urea by urease to give carbon dioxide and ammonium ions. The latter causes a change in colour of a pH sensitive dye.

Further reading: Kutter, D. (1977). *Rapid Clinical Diagnostic Tests.* (Munich–Vienna–Baltimore: Urban and Schwarzenberg)

B

BALANCE STUDIES

Procedures which can be used to give an indication of the status of certain substances within the body, e.g. fat, calcium, and nitrogen. Balance studies involve measuring the intake of the substance and comparing it with the amount lost in faeces and urine. Positive balances occur when intake exceeds loss, for example nitrogen and calcium in pregnancy. Negative balances are found when loss exceeds intake, for example in malabsorption.

See also: **faecal fat**

BANTU SIDEROSIS

A form of iron overloading seen among the African population of Southern Africa. The excess iron intake comes mainly from beer which is often brewed in iron containers.

See also: **haemochromatosis**

BARBITURATES

A family of drugs which have the basic formula

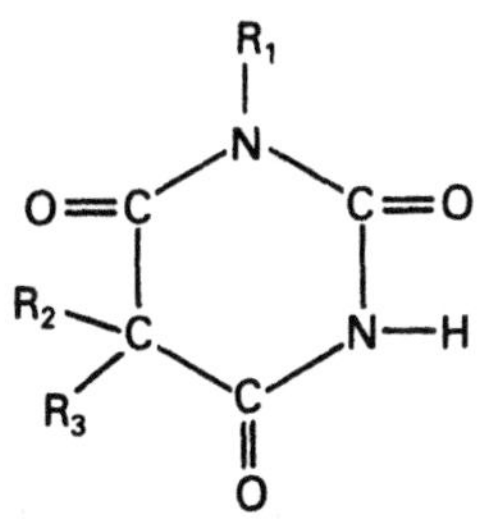

where R = various hydrocarbon groups, e.g. ethyl, phenyl etc.

Since they act as central nervous system depressants, they have been prescribed extensively as sedatives although their use is now being discouraged. Barbiturates are classified into four types, depending on the duration of their pharmacological action:

(1) Long-acting, e.g. phenobarbitone

(2) Intermediate-acting, e.g. butobarbitone.

(3) Short-acting, e.g. cyclobarbitone.

(4) Ultrashort-acting, e.g. thiopentone. These are used as anaesthetic agents and are therefore rarely encountered in cases of overdosage.

Measurement

(1) Spectrophotometric technique of Broughton. In this method barbiturates are extracted into chloroform and then re-extracted back into alkali solution. The extract is scanned at two different pH's, when the barbiturate molecule has different absorption spectra due to its existence in two different ionized forms. The difference between the optical densities of the two forms at 260 nm can be used to calculate the amount of barbiturate present.

(2) Gas–liquid chromatographic or HPLC methods.

(3) Antibodies have been raised against barbiturates enabling them to be assayed by radioimmunoassay or enzyme-immunoassay.

Identification of an unknown barbiturate

(1) Heat treatment of the alkaline extract in the Broughton procedure. Short-acting barbiturates are more stable to this treatment than intermediate- or long-acting types.

(2) Chromatographic procedures, e.g. thin-layer or gas–liquid chromatography.

Further reading: Yeoman, W.B. (1971). Toxicological analysis in the clinical chemistry laboratory. *Ann. Clin. Biochem.*, **8**, 93

Meade, B.W. *et al.* (1972). Technical Bulletin No. 24. Simple tests to detect poisons. *Ann. Clin. Biochem.*, **9**, 35

BARRIER LAYER CELL

A device used in some spectrophotometers for measuring the intensity of light. It consists of a thin layer of silver on a layer of the semiconductor selenium. These are mounted onto an iron backing, which because it is deficient in electrons acts as a positive electrode. The silver layer acts as a negatively charged electrode. When light falls on the silver layer, electrons flow from the selenium into the iron backing and this can be detected by means of a galvanometer. The electron flow is directly proportional to the intensity of the light falling on the silver layer.

Further reading: General list of analytical textbooks

BARTTER'S SYNDROME

A rare disorder in which there is hypokalaemia, hyperaldosteronism and hyperreninaemia but with normal blood pressure. It is possible that overproduction of renal prostaglandins may be a factor in the pathogenesis of the disease.

Further reading: Editorial (1976). Bartter's syndrome. *Lancet*, **2**, 721

BASAL METABOLIC RATE

An indication of the rate of metabolism in the body which consists of measuring the O_2 intake and CO_2 output. It can be used to assess the thyroid status of the body, since thyroid hormones influence the rate of many metabolic processes.

See also: **thyroxine**

BASE DEFICIT

See: **base excess**

BASE EXCESS

This is defined as the number of millimoles of acid required to titrate one litre of blood to a pH of 7.4 at a PCO_2 of 5.3kPa (40 mmHg) at 37 °C. It is usually determined indirectly. Positive values indicate a relative deficit of non-carbonic acid while negative values indicate a relative excess of non-carbonic acid in the blood. Negative values of the base excess are sometimes referred to as the base deficit.

See also: **Astrup**

BEER'S LAW

The amount of light absorbed by a solution is proportional to the number of absorbing molecules in it.

See also: **absorption**

BENCE-JONES PROTEIN

A protein found in the urine of many patients with immunocytomas. They are monoclonal light chains or fragments of light chains which have been produced in amounts greater than those of the heavy chains. Because of their low molecular weight they pass through the glomerulus and are excreted in the urine. Bence-Jones proteins can form casts in the renal tubules and may also be involved in amyloid deposition in tissues. Bence-Jones proteinuria may be detected by Bradshaw's test or by Osgood–Haskin's test. Immunoelectrophoresis of the concentrated urine is however the most reliable method for its detection.

See also: **myeloma**

Further reading: Hobbs, J.R. (1975).Bence-Jones proteins. In Marks, V. and Hales, C.N., (Eds.) *Essays in Medical Biochemistry,* Vol. 1., p. 105. (London: The Biochemical Society and the Association of Clinical Biochemists)

BENEDICT'S TEST

A test for the presence of reducing substances in urine. Benedict's reagent contains cupric ion complexed with citrate in alkaline solution. Glucose or other reducing substances reduce cupric ion to cuprous ion, resulting in the formation of yellow cuprous hydroxide or red cuprous oxide depending on the amount of reducing substance present.

See also: **reducing substances in urine**

BENIGN PARAPROTEINAEMIA

The term used to describe the situation in which a paraprotein has been found in the blood of a patient who has no evidence of a malignant condition such as myeloma or macroglobulinaemia. A benign paraproteinaemia is indicated if the paraprotein level fails to rise over a period of years.

See also: **paraprotein**

BENZODIAZEPINES

A family of drugs which are widely prescribed as tranquilizers and sleeping pills. Examples include diazepam (Valium), nitrazepam (Mogadon) and chlordiazepoxide (Librium). They are frequently taken in suicide attempts and can be detected in biological fluids by a number of techniques including examination of their absorption spectra following extraction. Diazotization reactions can also be used for their detection.

Further reading: Clifford, J.M. and Smyth, W.F. (1974). The determination of some 1,4-benzodiazepines and their metabolites in body fluids. A review. *The Analyst*, **99**, 1178

BERTHELOT REACTION

In the presence of the catalyst nitroprusside, ammonia reacts with hypochlorite and phenol to give indophenol, which is coloured blue in alkaline solution. This reaction can therefore be used to estimate blood ammonia. It can also be used to measure urea when used in conjunction with the enzyme urease which catalyses the conversion of urea to ammonia.

See also: **ammonia, urea**

BIAL'S TEST

A test which can be used for the detection of urinary pentoses. When heated with hydrochloric acid, pentoses are converted to furfural which reacts with orcinol to form green coloured compounds.

BIAS

A factor which causes a variation in the value of a parameter from what it should actually be. This may be caused by for example a particular method, a particular instrument or a particular operator. The term can also be used in the same sense as accuracy.

See also: **quality control**

BICARBONATE

An important component of acid–base homeostasis. It exists in equilibrium with carbonic acid and carbon dioxide as follows:

$$CO_2 + H_2O \rightleftharpoons H_2CO_3 \rightleftharpoons H^+ + HCO_3^-$$

When the Henderson–Hasselbalch equation is applied to these reactions, bicarbonate can be related to pH as follows:

$$pH \propto \frac{[HCO_3^-]}{[H_2CO_3]} \text{ or } \frac{[HCO_3^-]}{[CO_2]}$$

$[HCO_3^-]$ can be regarded as the metabolic component of the acid–base picture while the $[CO_2]$ or PCO_2 can be regarded as the respiratory component. Thus a change in one of these components may result in a compensatory change in the other component in order to maintain the pH within normal limits.

Increases in plasma bicarbonate

(1) In compensated respiratory acidoses such as chronic obstructive airway diseases. In these conditions, the PCO_2 increases and this can be followed by a compensatory rise in bicarbonate to maintain the pH within normal limits.

(2) In metabolic alkaloses, e.g. excessive alkali ingestion or potassium depletion.

Decreases in plasma bicarbonate

(1) In compensated respiratory alkaloses that result from hyperventilation. In these conditions, the PCO_2 decreases and this is followed by a compensatory fall in bicarbonate, thus maintaining the pH within normal limits.

(2) In metabolic acidoses, e.g. diabetic ketoacidosis and renal failure.

Measurement

Bicarbonate in blood is difficult to determine as such. There are several different approaches to its determination:

(1) Methods in which bicarbonate is liberated as carbon dioxide. When plasma is treated with acid, carbon dioxide is liberated. This carbon dioxide is derived from dissolved CO_2, carbonic acid, and bicarbonate (see above equation) and is known as total CO_2. The CO_2 liberated can be measured by one of several different techniques:

(a) Gasometric methods. The volume of gas liberated can be measured using equipment such as the Van Slyke apparatus or the Natelson microgasometer. In some methods a small correction may be applied to allow for the relatively small fraction of dissolved CO_2 and carbonic acid.

A variation on the gasometric type of analysis is the 'alkali reserve' or the 'CO_2 combining power'. This is the plasma CO_2 content after equilibration of the plasma at a PCO_2 of 5.3 kPa (40 mmHg). It is an attempt to replace the CO_2 which has been lost during the collection of the specimen with the alveolar air of a normal person.

(b) Titrimetric methods. In these techniques the excess acid added to the sample is back-titrated with standard alkali.

(c) Continuous flow methods. The plasma sample is diluted in acid. The liberated CO_2 is separated by a trap and is used to segment a stream of buffered phenolphthalein. The change in the intensity of the red colour of the phenolphthalein depends upon the pH and therefore the CO_2 content. Sodium carbonate solutions are used as standards.

(2) PEP carboxylase methods. Treatment of the plasma with an alkaline buffer results in the conversion of the dissolved CO_2, bicarbonate and carbonic acid to the bicarbonate form. The bicarbonate is made to react with a phosphoenolpyruvate (PEP) carboxylase/malate dehydrogenase coupled enzyme system as follows:

$$\text{phosphoenolpyruvate} + \text{bicarbonate} \xrightarrow{\text{PEP carboxylase}} \text{oxaloacetate} + \text{phosphate}$$

$$\text{oxaloacetate} + \text{NADH} \xrightarrow{\text{malate dehydrogenase}} \text{malate} + \text{NAD}$$

The change in the absorbance at 340 nm, due to the disappearance of NADH, is directly proportional to the bicarbonate concentration. This method is used on several discrete analysers.

(3) Standard bicarbonate. This is the concentration of bicarbonate in whole blood at 37 °C equilibrated at a PCO_2 of 5.3 kPa (40 mmHg) with the blood haemoglobin fully oxygenated. It can be derived if pH measurements are taken of two portions of the blood specimen equilibrated with two different gas mixtures of different PCO_2 content. See **Astrup** for details of the measurement and **standard bicarbonate** for explanation of the usefulness of its determination.

(4) Several of the modern automatic blood gas analysers calculate both 'actual' and standard bicarbonate from measurements of pH and PCO_2, using the Henderson–Hasselbalch equation.

BILE

A secretion of the liver which is stored in the gall bladder. Bile contains among other things bile salts, conjugated bilirubin, cholesterol and phospholipids. It eventually passes into the intestine where the bile salts assist digestion by emulsifying fats.

See also: **bile acids and salts**

BILE ACIDS AND SALTS

Two primary bile acids are synthesized in the liver from cholesterol. These are cholic acid and chenodeoxycholic acid. They are conjugated with the amino acids, glycine and taurine, to form bile salts (for example taurine may be conjugated with cholic acid to give taurocholic acid). The bile salts are excreted in the bile, and in the gut they are acted upon by bacteria which convert them to the secondary bile acids, deoxycholic acid and lithocholic acid. Some of the secondary bile acids are absorbed and carried by the enterohepatic circulation to the liver where they are re-excreted.

Function

Bile salts possess both polar and non-polar ends and are thus able to coalesce into micelles with the polar (water soluble) groups on the outside and the non-polar groups in the centre forming a lipid soluble environment. In this way they are able to assist in the emulsification and absorption of fat.

Pathology

Increased levels of bile acids in the serum are found in hepatitis and obstructive jaundice. In cholestatic liver disease, there can be a deficiency of bile salts in the intestine and this leads to fat malabsorption. Disturbances in bile acid composition may be one of the factors involved in gallstone formation because it may lead to conditions favouring cholesterol precipitation.

Measurement

Techniques such as fluorimetry or gas–liquid chromatography are required to measure bile acids in the serum. Excess bile acids in the urine can be detected by the Hay test or the Pettenkofer reaction. In the Hay test, flowers of sulphur are dropped onto the urine. If excess bile acids are present, the surface tension is lowered and the particles sink, whereas with normal urine, they float. The Pettenkofer test is the reaction of bile acids with fructose or furfural to form a red colour in the presence of sulphuric acid.

Further reading: General list of clinical textbooks

BILE PIGMENTS

The term used to describe bilirubin and its breakdown products.

See: **bilirubin**

BILIARY CIRRHOSIS

See: **primary biliary cirrhosis**

BILIRUBIN

Metabolism

Red blood cells are broken down by the reticulo-endothelial system and the released haemoglobin is split into haem and the protein, globin. The haem molecule is then split to give biliverdin which is rapidly converted to bilirubin. The bilirubin is carried to the liver bound to plasma albumin. At this stage it is unconjugated and because it is water insoluble it does not appear in the urine

The bilirubin enters the liver cells where it is conjugated with glucuronic acid by means of the glucuronyl transferase system to form water soluble diglucuronide derivatives. These are excreted in the bile and pass into the gut. In the colon, bacteria reduce bilirubin to mesobilirubinogen, urobilinogen and stercobilinogen, this group of pigments being known collectively as faecal urobilinogen. A portion of this is absorbed into the circulation and from this a small proportion is excreted in the urine. The majority however is re-excreted in the bile where it can be oxidized to urobilin.

Disturbances of bilirubin metabolism

Increases in plasma bilirubin cause jaundice and this may arise because of several categories of disorder:

(1) There may be increased bilirubin production, e.g. as a result of haemolysis. In this type of jaundice, the serum bilirubin is of the unconjugated variety.

(2) There may be defective uptake of bilirubin by the liver cell as in Gilbert's disease. Again, the bilirubin is of the unconjugated variety.

(3) There may be failure of the liver to conjugate bilirubin as in physiological neonatal jaundice. Another condition where this occurs is Crigler–Najjar syndrome where there is a deficiency of the glucuronyl transferase system. In these conditions, the bilirubin is mainly of the unconjugated variety.

(4) There may be a failure to excrete the conjugated bilirubin and this may occur in cholestatic conditions, (e.g. gallstones, biliary cirrhosis) and the congenital Dubin–Johnson syndrome. In this group of diseases there is a rise in conjugated as well as unconjugated bilirubin.

Measurement of bilirubin in serum

There are two approaches to bilirubin quantitation in serum:

(1) *Spectrophotometric* Bilirubin absorbs light at 461 nm, the amount of light absorbed being proportional to its concentration. However, haemoglobin, which may be present as a result of haemolysis, also absorbs light at 461 nm so a correction has to be made for this. This is done by measur-

ing the absorbance of the sample at 551 nm where haemoglobin has the same absorbance as it does at 461 nm. Subtraction of the absorbance at 551 nm from the absorbance at 461 nm gives the corrected absorbance due to bilirubin alone. This can be related to a standard curve and the bilirubin content determined. There are several instruments on the market which measure simultaneously the absorbances at these two wavelengths and give a direct readout of the bilirubin concentration.

(2) *Diazo methods* Many laboratories estimate bilirubin by its reaction with diazotized sulphanilic acid to form azobilirubin (the Van den Bergh reaction). Conjugated bilirubin reacts rapidly and is therefore sometimes referred to as direct-acting bilirubin. On the other hand, unconjugated bilirubin will only react if an accelerating agent is present and this is referred to as the indirect reaction.

Many modifications of the Van den Bergh reaction have been tried, using different accelerators. Among these accelerators are methanol (Malloy and Evelyn), ethanol (King and Coxon), benzoate and urea (Powell), diphylline (Michaelson) and caffeine/benzoate (Jendrassik and Grof). This latter method is one of the most widely used. In this technique the diazotization reaction is terminated by the addition of ascorbic acid which destroys the excess diazo reagent. Alkali is added at the end of the reaction to give a more stable colour and also to dissolve any proteins that may have been precipitated.

As well as diazotized sulphanilic acid, other diazo reagents have been tried, e.g. 2,4-dichloroaniline.

Bilirubin standardization has always been a problem because of its instability. Bilirubin can be purchased and used as a standard, either in chloroform or added to serum. Alternatively some laboratories use commercially assayed sera with assigned values. Another alternative to bilirubin standardization is to use another more stable chromogen, e.g. methyl red.

Detection of bilirubin in urine

Bilirubin in urine can be detected by a diazotization reaction (e.g. the Ictotest marketed by Ames) or by its reaction with ferric chloride (Fouchet's test).

Further reading: General list of analytical and clinical textbooks

BILIVERDIN

A green pigment formed from haemoglobin following the degradation of red blood cells. It is rapidly converted to bilirubin.

See also: **bilirubin**

BIOTIN

A water-soluble vitamin of the B group. It is an essential coenzyme for carboxylase reactions in man. Dietary deficiency is rare but when it occurs produces the symptoms of alopecia and dermatitis.

Further reading: General list of clinical textbooks

BISALBUMINAEMIA

A rare inborn error in which the albumin of an affected individual separates into two distinct peaks during serum protein electrophoresis. There do not appear to be any pathological consequences of the condition.

Further reading: Bearn, A.G. and Litwin, S.D. (1978). Deficiencies of circulating enzymes and plasma proteins. In Stanbury J.B., Wyngaarden, J.B. and Fredrickson, D.S. (eds.) *The Metabolic Basis of Inherited Disease,* 4th Edn., p. 1712. (New York: McGraw-Hill.)

BIURET REACTION

A reaction between the peptide bonds of proteins and cupric ions in alkaline solution to form a coloured chelated complex of unknown composition. Because an analagous reaction takes place between cupric ions and the organic compound biuret (NH_2-CO-NH-CO-NH_2) the reaction is referred to as the biuret reaction. The intensity of the colour produced is proportional to the number of peptide bonds reacting and it can therefore be used as a method of protein quantitation.

See also: **proteins**

BLIND LOOP SYNDROME

A condition in which there is stagnation of intestinal contents which aids bacterial multiplication in the small intestine. The bacteria can metabolize bile salts and this results in fat malabsorption and steatorrhoea. Vitamin B_{12} deficiency may also occur since many bacteria take up the available vitamin B_{12}. Blind loop syndrome can occur as a result of surgery or in the presence of diverticula.

Further reading: General list of clinical textbooks

BLOOD GASES

Strictly speaking the levels of oxygen and carbon dioxide in the blood. In practice, a full blood gas analysis also involves measuring other acid–base parameters in addition, e.g. pH and bicarbonate.

See also: **bicarbonate, carbon dioxide, oxygen, pH**

BLOOD UREA NITROGEN (BUN)

A method of expressing urea concentrations favoured in North America. It enables the amount of nitrogen in urea to be compared with that of other non-protein nitrogen components in body fluids.

See also: **urea**

BLOOD VOLUME

In normal subjects blood volume is approximately 80 ml/kg body weight. It can be determined by measuring the dilution of an injection of radioactively labelled red cells. Alternatively plasma volume may be determined using radioactively labelled albumin and if the packed cell volume is known, the blood volume may be calculated.

BODANSKY UNIT

One of the units by which alkaline phosphatase activity can be expressed. It is defined as the enzyme activity in 100 ml of serum which will liberate 1 mg of phosphorus from β-glycerophosphate at pH 8.6 in one hour at 37 °C.

See also: **alkaline phosphatase**

BODY SURFACE AREA

This must be taken into consideration in renal studies, since clearance is proportional to body surface area. It can be calculated from the weight and height of the individual.

BODY WATER

A 70 kg man contains about 45 litres of water, of which about 45% is extracellular and 55% intracellular. Normally water intake and output are balanced, the amount lost in urine, faeces, sweat and expired air being replaced by water taken in food and drink.
See also: **water balance**

BOHR EFFECT

An alteration in the affinity of haemoglobin for oxygen. Decreases in blood pH or increases in PCO_2 cause the haemoglobin dissociation curve to move to the right. Alternatively an increase in blood pH or a decrease in PCO_2 cause the curve to move to the left.
See also: **haemoglobin,** P_{50}.

BONE

A tissue consisting of a collagenous protein matrix impregnated with mineral salts especially hydroxyapatite (a form of calcium phosphate). Cells called osteoblasts are involved in the formation of the collagen fibre network. Also present in bone are cells called osteoclasts which are responsible for bone resorption.

Osteoblasts contain high levels of alkaline phosphatase and this enzyme may be involved in the hydrolysis of phosphate esters, leading to calcium phosphate deposition.

See also: **calcium**

BORON

Boron is found in boric acid and borate salts which are used as mild antiseptics or for laundering. Occasionally cases of poisoning may be encountered. Borates can be detected in the urine of such subjects by their reaction with turmeric yellow to give a red-brown colour.

Further reading: General list of analytical textbooks

BOVRIL TEST

A test for growth hormone deficiency based on the fact that growth hormone secretion normally increases in response to amino acids. In this test, amino acids in the form of Bovril are given orally and samples of blood are taken for growth hormone estimation. Failure of growth hormone levels to rise significantly are an indication of possible growth hormone deficiency. This test is particularly useful as a screening test in children.

See also: **human growth hormone**

BRADSHAW'S TEST

A test for Bence Jones protein in urine. A few millilitres of urine are carefully placed onto concentrated hydrochloric acid. If Bence Jones protein is present, a layer of turbidity appears at the interface. The test has a false-negative rate of about 5%. False positives can occur if other proteins are present in large amounts, e.g. as in nephrotic syndrome.

See also: **Bence Jones protein**

BROMIDES

Bromide drugs, in either organic or inorganic form, are used as sedatives. Bromide in serum can be measured by its reaction with gold trichloride, when it displaces the chloride resulting in the formation of gold tribromide which can be measured spectrophotometrically.

Further reading: General list of analytical textbooks

BROMOCRESOL GREEN

A dye which can bind to albumin, resulting in a change in the spectral characteristics of the dye. This can be used as the basis of a method for determination of serum albumin.

See also: **albumin**

BROMOCRIPTINE

A drug which inhibits prolactin secretion by its action on the pituitary. It is used clinically to suppress puerperal lactation and in the treatment of hyperprolactinaemic hypogonadism. It also affects growth hormone release and has been used in treating acromegaly.

Further reading: Lees, A.J. (1977). Bromocriptine. *Br. J. Hosp. Med.*, **18**, 336

Jacobs, H.S. and Wright, C.S. (1978). Bromocriptine in obstetrics and gynaecology. *Br. J. Hosp. Med.*, **20**, 652

Lewis, M.J. (1978). Bromocriptine in hypertension. *Br. J. Hosp. Med.*, **20**, 661

Stern, G. and Lees, A. (1978). Bromocriptine in Parkinson's disease. *Br. J. Hosp. Med.*, **20**, 666

Harrower, A.D.B. (1978). Bromocriptine in anorexia nervosa. *Br. J. Hosp. Med.*, **20**, 672

BROMSULPHTHALEIN TEST

A test that can be used to assess liver function when other liver function tests are normal. It is a test of the ability of the liver to conjugate and excrete the dye bromsulphthalein (BSP). It consists of an intravenous injection of the dye followed by the collection of blood specimens up to 45 minutes after the injection. Bromsulphthalein is then measured in the serum samples by adding alkali which renders the dye purple. In normal subjects less than 5% of the dose should remain in the circulation after 45 minutes. Increased retention of BSP is a sensitive index of hepatic dysfunction although false positives may occur if there is impaired circulation in the liver.

See also: **liver function tests**

BRUTON'S DISEASE

See: **sex-linked agammaglobulinaemia**

BUFFY LAYER

This blood fraction contains white blood cells and platelets. Ascorbic acid (vitamin C) is carried mainly in the white blood cells and so its measurement in the buffy layer gives an indication of the vitamin C status of the body.

See also: **ascorbic acid**

BUTANOL-EXTRACTABLE IODINE

A method of thyroid hormone estimation which consists of extracting iodinated amino acids from serum using n-butanol. The extract is washed with alkali which removes inorganic iodine and mono- and di-iodotyrosines, leaving behind tri-iodothyronine and thyroxine. The butanol extract can then by evaporated to dryness and analysed for iodide content.

See also: **thyroxine**

C

C3 (βC -GLOBULIN)

The component of the complement system which is present in the serum in the greatest amount, enabling it to be measured by routine immunochemical techniques such as radial immunodiffusion. C3 levels are often measured as an indication of the overall activity of the complement pathway.

It behaves electrophoretically as a β-globulin. Upon ageing in serum it breaks down into two fragments, β_1A and α_2D. These fragments of C3 still have antigenic determinants which enable them to react with antiserum raised against C3. Hence C3 levels are often expressed in terms of β_1C/β_1A-globulin.

Low serum levels

These are usually indicative of increased complement utilization and are thus found in those diseases which have an immune element.

(1) In glomerulonephritis when autoantibodies or immune complexes are adsorbed onto the basement membrane activating complement and in this way damaging the membrane.

(2) Similar mechanisms probably occur in the renal complications of systemic lupus erythematosus, when immune complexes can be deposited in the glomeruli.

(3) In acquired haemolytic anaemias.

(4) In liver diseases (due to decreased synthesis).

High serum levels

These occur in acute inflammatory conditions.

See also: **complement**

C4 (β_1E-GLOBULIN)

One of the nine major components of the complement system. Apart from C3, it is the complement component present in the greatest amounts in serum, and therefore, along with C3, it can be measured as an indication of the overall activity of the complement system. Immunochemical techniques (e.g. radial immunodiffusion) are used for its assay.

See also: **complement**

CADMIUM

Exposure to cadmium can cause renal tubular damage resulting in an aminoaciduria. The element itself is usually measured by atomic absorption spectrophotometry.

Further reading: Special Issue. (1975). Trace elements in clinical chemistry. *Clin. Chem.*, **21** (4)

CAERULOPLASMIN

A plasma protein having a molecular weight of approximately 160 000 and having α_2-globulin mobility. It is involved in copper transport and contains more than 90% of the plasma copper. Decreased plasma levels are found in Wilson's disease.

Measurement

(1) Caeruloplasmin behaves as an oxidase and can be measured by its enzymic activity using *p*-phenylenediamine as a substrate.

(2) It can also be determined immunochemically by e.g. radial immunodiffusion or the Laurell "rocket" technique.

See also: **Wilson's disease**

Further reading: Gutteridge, J.M.C. (1978). Caeruloplasmin: a plasma protein, enzyme and antioxidant. *Ann. Clin. Biochem.,* **15**, 293

CAFFEINE STIMULATION TEST

A test of the ability of the stomach to secrete hydrochloric acid in response to caffeine which is introduced into the stomach by means of a tube. In normal subjects caffeine stimulates gastric acid secretion.

See also: **Diagnex Blue test**

CALCIFEROL

See: **vitamin D**

CALCITONIN

A polypeptide hormone, consisting of 32 amino acids, produced in the C cells of the thyroid gland. It is involved in calcium homeostasis, its main action being to decrease osteoclastic activity and therefore reduce bone resorption. Calcitonin also has a phosphaturic effect. It is secreted in response to high ionized calcium levels. High serum levels are found in patients with medullary carcinoma of the thyroid. It is usually measured by radioimmunoassay.

See also: **calcium**

CALCIUM

Role

More than 99% of the body's calcium is present in bone as calcium fluorophosphate apatite. Calcium is also involved in neuromuscular excitability and nervous impulse transmission, blood coagulation and in the transfer of inorganic ions across cell membranes.

Intake

Calcium is absorbed in the upper small intestine, a process which is promoted by vitamin D. Optimal absorption occurs at acid pHs. High dietary concentrations of phosphate, phytate or oxalate decrease absorption because of complex formation.

Serum calcium

Calcium in serum is present in three distinct forms:

(1) The non-diffusible protein-bound form.

(2) Complexed with certain ions, e.g. citrate, phosphate.

(3) Ionized calcium. This is the physiologically active form.

Control of serum calcium levels

Three hormonal mechanisms are involved in calcium homeostasis.

(1) Parathyroid hormone (PTH) raises the serum ionized calcium level by three types of action.

(a) It acts directly on bone osteoclasts, releasing bone salts.

(b) It causes decreased phosphate reabsorption in the tubules. This decreases serum phosphate and therefore by a mass action effect causes release of phosphate salts (and therefore calcium) from bone.

(c) It increases the synthesis of 1,25-dihydroxycholecalciferol which increases calcium absorption from the gut.

(2) Vitamin D increases the intestinal absorption of calcium.

(3) Calcitonin, produced by the thyroid, reduces plasma calcium levels by decreasing osteoclastic activity.

Hypercalcaemia

Hypercalcaemia can result in:

(1) Deposition of calcium in the kidney leading to renal failure.

(2) Decreased neuromuscular excitability.

(3) Cardiac arrest.

Among the more common causes of hypercalcaemia are:

(1) Primary hyperparathyroidism e.g. parathyroid adenoma.

(2) Tertiary hyperparathyroidism. This is when the parathyroid gland has been under prolonged stimulation by low serum ionized calcium with the result that when the cause of the hypocalcaemia has been removed, hypersecretion of PTH has become autonomous.

(3) Ectopic production of PTH.

(4) Vitamin D overdosage.

(5) Sarcoidosis.

(6) Myeloma.

(7) Thyrotoxicosis.

(8) Certain cases of breast carcinoma (thought to be due to the production of a substance with vitamin D like properties)

Hypocalcaemia

Low ionized calcium levels result in tetany, and prolonged hypocalcaemia causes cataracts and mental symptoms. Among the more common causes of hypocalcaemia are-

(1) Hypoparathyroidism and pseudohypoparathyroidism.

(2) Calcium and vitamin D deficiency.

(3) Renal failure.

Urinary calcium

If renal function is normal, any form of hypercalcaemia causes hypercalcuria, although hypercalcuria without hypercalcaemia is found in the not uncommon condition of idiopathic hypercalcuria.

Measurement of calcium in body fluids

(1) Precipitation techniques. Calcium can be precipitated as calcium oxalate and the oxalate can be measured by its reaction with potassium permanganate.

In another technique calcium can be precipitated as calcium chloranilate which can be estimated colorimetrically.

(2) Titration methods. A diluted sample of the fluid is titrated with EDTA in the presence of an indicator dye (e.g. calcein, murexide, Eriochrome Black T) which is bound as a complex to calcium. When all the calcium present has been chelated by EDTA the spectral characteristics of the dye are changed and this can be detected fluorimetrically or colorimetrically. Magnesium interference in some of these methods can be eliminated by titration at an alkaline pH at which magnesium is precipitated as magnesium hydroxide.

(3) Calcium can be estimated by its formation with cresolphthalein complexone to form a coloured complex. 8-Hydroxyquinoline can be added to bind magnesium and prevent interference from this ion. Other complexing agents have been tried, e.g. methylthymol blue. This type of method is used in many continuous flow and discrete analysers.

(4) Atomic absorption methods. The main problem with these methods is that anions such as phosphates bind to calcium and these compounds do not dissociate in the flame,

causing falsely low results. This interference can be overcome by adding lanthanum which preferentially binds with phosphate and prevents the formation of calcium phosphate.

Ionized calcium

This is the form of calcium which is physiologically active in the serum. There are several approaches to its determination.

(1) By calculation. There are several formulae which can be used to calculate ionized calcium from the total serum calcium and the protein level.

(2) By ion-selective electrodes.

(3) By ultrafiltration or ultracentrifugation.

Further reading: Wills, M.R. (1974). Hypercalcaemia. *Br. J. Hosp. Med.*, **11**, 279

Editorial (1977). Correcting the calcium. *Br. Med. J.*, **1**, 598

Editorial (1977). Dietary calcium. *Br. Med. J.*, **2**, 1105

Editorial (1979). Serum calcium. *Lancet*, **1**, 858

CALCIUM PYROPHOSPHATE

Precipitation of calcium pyrophosphate crystals in joint cavities causes pseudogout which is clinically similar to gout. Calcium pyrophosphate crystals can be distinguished from uric acid crystals using a polarizing microscope.

Further reading: Scott, J.T. (1975). The analysis of joint fluids. *Br. J. Hosp. Med.*, **14**, 653

CALCULUS

This is a stone which may be formed in secreting organs or their ducts. They have been found in the salivary gland, pancreas and prostate but are most frequently encountered in the urinary tract and gall bladder. Urinary calculi usually consist of calcium, magnesium, oxalate, carbonate or phosphate but occasionally uric acid, cystine and xanthine stones occur as a result of a metabolic disease. Cholesterol and bilirubin are found in biliary stones.

CARBAMYLPHOSPHATE SYNTHETASE DEFICIENCY

A rare inborn error of metabolism in which there is a deficiency of this urea cycle enzyme. Hyperammonaemia is a feature of the disease.

Further reading: Shih, V. (1978). Urea cycle disorders and other congenital hyperammonaemic syndromes. In Stanbury, J.B., Wyngaarden, J.B. and Fredrickson, D.S. (eds.) *The Metabolic Basis of Inherited Disease*, 4th Edn., p. 362. (New York: McGraw-Hill)

CARBON DIOXIDE

This is formed in the tissues from the oxidation of fats and carbohydrates. It is eventually excreted in the lungs. In the blood it exists in equilibrium with carbonic acid and bicarbonate as follows:

$$CO_2 + H_2O \rightleftharpoons H_2CO_3 \rightleftharpoons H^+ + HCO_3^-$$

From the Henderson–Hasselbalch equation, the pH of the blood can be expressed as:

$$\mathrm{pH} \propto \frac{\text{bicarbonate concentration}}{CO_2 \text{ concentration}}$$

The concentration of CO_2 in blood is usually represented as PCO_2, the partial pressure of CO_2. Bicarbonate concentration can be regarded as the metabolic component of acid–base homeostasis while CO_2 can be regarded as the respiratory component. Thus a primary change in one of these components as a result of a clinical condition can result in a compensatory change in other component. Raised blood PCO_2 occurs in respiratory acidosis (as in chronic obstructive airway disease) and compensated metabolic alkalosis. Low blood PCO_2 is found in respiratory alkalosis (hyperventilation) and in compensated metabolic acidosis.

Measurement of PCO_2

(1) PCO_2 can be measured directly by a PCO_2 electrode (the Severinghaus electrode). This is essentially a pH electrode in contact with a bicarbonate solution. It is separated from the sample by a membrane which is permeable to gas but not to solutions. CO_2 diffuses across the membrane from the sample into the bicarbonate solution. This results in a

change in the pH, caused by a re-equilibration of the components in the Henderson–Hasselbalch equation. This pH change can be detected by the electrode.

(2) The PCO_2 can be determined by equilibration methods in which the blood is equilibrated with two gases with different, but known, PCO_2 values. For more details of this technique see under **Astrup.**

Further reading: Davenport, H.W. (1974). *The ABC of Acid–Base Chemistry.* 6th Edn. (Chigago: The University of Chicago Press)

Siggaard-Anderson, O. (1974). *The Acid–Base Status of the Blood.* 4th Edn. (Copenhagen: Munksgaard)

CARBON DIOXIDE COMBINING POWER

See: **alkali reserve**

CARBONIC ANHYDRASE (CARBONATE DEHYDRATASE)

An enzyme which catalyses the interconversion of CO_2 and carbonic acid as follows:

$$CO_2 + H_2O \rightleftharpoons H_2CO_3$$

It is particularly important in three respects.

(1) In the gastric acid secreting cells of the stomach where it is involved in the secretion of hydrochloric acid.

(2) In the red blood cells. At the tissues, where CO_2 is being produced, it catalyses the formation of H_2CO_3 from H_2O and CO_2. At the lungs, the reverse occurs and this assists in the elimination of CO_2.

(3) In the renal tubular cells where it catalyses the formation of H_2CO_3 from H_2O and CO_2. The H_2CO_3 dissociates to give H^+ and HCO_3^-. H^+ is secreted in the urine, while the bicarbonate is retained.

Further reading: General list of clinical textbooks

CARBON MONOXIDE

See: **carboxyhaemoglobin**

CARBOXYHAEMOGLOBIN

A compound formed by the combination of carbon monoxide with haemoglobin. Carboxyhaemoglobin is more stable than oxyhaemoglobin (i.e. gives up carbon monoxide less easily than oxygen) and, as a consequence, tends to accumulate in the blood, resulting in tissue anoxia. The absorption spectrum of carboxyhaemoglobin is different from that of oxyhaemoglobin, the former being more pink in colour. These different spectral characteristics, can be used as a means of identifying and measuring carboxyhaemoglobin levels.

Measurement

(1) Carboxyhaemoglobin can be estimated by its characteristic absorption spectrum using spectrophotometric or spectrographic techniques.

(2) It can be measured indirectly by releasing carbon monoxide from the haemoglobin complex and measuring the gas, e.g. by gasometric techniques, gas chromatography, infra red spectroscopy or microdiffusion using a Conway unit. In this last technique, the reduction of palladium chloride by carbon monoxide to metallic palladium is estimated colorimetrically.

Further reading: Yeoman, W.B. (1971). Toxicological analysis in the clinical chemistry laboratory. *Ann. Clin. Biochem.*, **8**, 93.

Meade, B.W., *et al.* (1972). Technical Bulletin No. 24. Simple tests to detect poisons. *Ann. Clin. Biochem.*, **9**, 35

CARBROMAL

A bromine-containing drug used in combination with pentobarbitone as a hypnotic. It can be detected in gastric aspirates by a simple screening test.

Further reading: Meade, B.W., *et al.* (1972). Technical Bulletin No. 24. Simple tests to detect poisons. *Ann. Clin. Biochem.*, **9**, 35

CARCINOEMBRYONIC ANTIGEN

A glycoprotein which is normally only present in the fetus. In adults it is produced by malignant tumours of the gastrointestinal tract, and therefore can be measured as an indicator of malignancy. However its usefulness may be limited as it is also produced in non-malignant diseases of the gastrointestinal tract such as ulcerative colitis and Crohn's disease. It can be measured by radioimmunoassay.

Further reading: Munro Neville, A. and Cooper, E.H. (1976). Biochemical monitoring of cancer. *Ann. Clin. Biochem.*, **13**, 283

CARCINOID SYNDROME

This is a tumour of argentaffin cells (cells which stain with silver salts), usually sited in the ileum or appendix, or less commonly in the bronchus, pancreas or stomach. These cells normally synthesize the hormone serotonin (5-hydroxytryptamine) which is a powerful smooth-muscle stimulant and vasoconstrictor. In carcinoid syndrome this hormone is secreted in excess amounts. The clinical symptoms of this condition include flushing, diarrhoea and bronchospasm. Carcinoid syndrome can be diagnosed biochemically by detection of the serotonin metabolite, 5-hydroxyindoleacetic acid, in excess amounts in the urine.

See also: **5-hydroxyindoleacetic acid, 5-hydroxytryptamine**

CARMINE

An inert dye which can be used as a faecal marker.

See also: **faecal markers**

CARNOSINAEMIA

A rare inborn error of metabolism in which there is a deficiency of the enzyme carnosinase which converts carnosine to β-alanine and histidine. Mental retardation is a clinical feature.

Further reading: Scriver, C.R., Nutzenadel, W. and Perry, T.L. (1978). Disorders of β-alanine and carnosine metabolism. In Stanbury, J.B., Wyngaarden, J.B. and Fredrickson, D.S. (eds.) *The Metabolic Basis of Inherited Disease,* 4th Edn., p. 528. (New York: McGraw-Hill)

β-CAROTENE

A fat-soluble substance, found in many plants, especially carrots. It can be hydrolysed in the intestine to give vitamin A.

Increases in serum levels

Increased serum carotene levels are found in some cases of hypothyroidism and when the dietary intake of carotenes is high.

Decreases in serum levels

Low serum carotene levels are found in steatorrhoea.

Measurement

(1) β-Carotene can be extracted from serum with petroleum ether and the yellow colour measured spectrophotometrically.

(2) β-Carotene can also be determined by its reaction with trifluoroacetic acid to produce a blue colour (Neeld–Pearson reaction).

See also: **vitamin A**

CAROTENE TOLERANCE TEST

A test of fat malabsorption which consists of measurement of serum carotene following the administration of an oral load of β-carotene to the patient. In normal subjects the serum carotenoid level rises, whereas in patients with malabsorption the serum level remains low.

See also: **vitamin A absorption test**

CARR–PRICE REACTION

The reaction of vitamin A with antimony trichloride in chloroform to produce a blue colour. It can be used for the estimation of vitamin A in serum

See also: **vitamin A**

CASTS

These are structures formed in the renal tubules whose shape they take. They eventually pass into the urine and can thus be examined microscopically. There are several varieties of cast and their detection can be of diagnostic significance. Hyaline casts for instance are found in all kidney diseases, whereas granular, epithelial and fatty casts suggest degenerative changes of the tubular epithelium.

CATECHOLAMINES

These are dihydroxy phenolic compounds having an amino group on the side chain. Examples of catecholamines are dihydroxyphenylethylamine (DOPamine) adrenaline and noradrenaline.

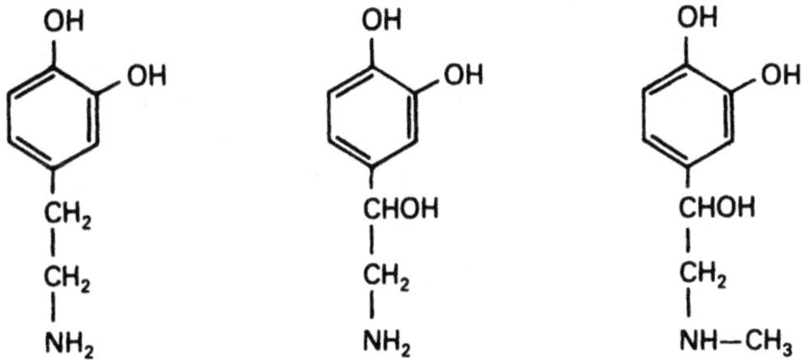

Dopamine Noradrenaline Adrenaline

Synthesis

Adrenaline and noradrenaline are synthesised from tyrosine as follows;

tyrosine ⟶ dihydroxyphenylalanine ⟶ dopamine ⟶ noradrenaline ⟶ adrenaline

Most noradrenaline is synthesised at the sympathetic nerve endings while adrenaline is mainly produced in the adrenal medulla.

Hypoglycaemia, cold, fear and injury stimulate their secretion.

Actions

Adrenaline and noradrenaline can cause variations in blood pressure by acting on the cardiovascular system. Adrenaline increases the rate of glycogenolysis, thus tending to raise the blood glucose level.

Metabolism

There are two principle mechanisms for the catabolism of adrenaline and noradrenaline;

(1) Oxidation, catalysed by the enzyme monoamine oxidase (MAO)

(2) Methylation of the hydroxyl groups, a reaction catalysed by catechol–O–methyl transferase.

The major metabolites resulting from these processes are the methylated derivatives metadrenaline and normetadrenaline, and 4-hydroxy, 3-methoxymandelic acid, (HMMA, VMA).

Increased catecholamine levels

Increased blood and urinary catecholamines and their metabolites are found in tumours of the sympathetic nervous system. There are two main types of tumour:

(1) Tumours of chromaffin tissue (phaeochromocytoma). These are mainly adrenal tumours.

(2) Tumours of nerve cells (neuroblastoma). These are mainly extra–adrenal.

Measurement of catecholamines

Although catecholamines can be measured in plasma, it is their levels in urine that are usually measured. Urinary adrenaline and noradrenaline can be measured by a fluorimetric technique after their isolation on an alumina column. However in most laboratories, it is catecholamine metabolites that are usually measured. These include:

(1) Metadrenaline and normetadrenaline. These can be isolated by column chromatography, converted to vanillin and then assayed spectrophotometricaly.

(2) 4-hydroxy, 3-methoxymandelic acid.

(3) Homovanillic acid. This is the major urinary metabolite of DOPA and DOPamine.

See also: **adrenaline, 4-hydroxy, 3-methoxymandelic acid, phaeochromacytoma**

Further reading: Catecholamines. *Br. Med. Bull.*, (1973), **22**, (2)

CELLULOSE ACETATE

A material which can be used as an electrophoretic support medium. It is supplied commercially as membranes, which contain a large proportion of air spaces. These air spaces become filled with liquid when wetted, transforming the membrane into a pliable sheet. One of the major advantages of cellulose acetate as an electrophoretic support medium is that adsorption is minimal, enabling sharply defined bands to be obtained. The sheets can be made transparent for densitometric measurements by the use of suitable solvents.

CENTRIFUGAL ANALYSERS

Automated instruments which use centrifugal force as the basis for the mixing of reagents. Reagents are placed in the inner compartments of a specially constructed centrifuge rotor. Acceleration of the rotor causes the liquids to move outwards to the outer compartment where mixing occurs. From here they pass to a cuvette. With rotation of the centrifuge, the cuvettes pass sequentially through a light beam when optical densities can be measured. Calculation of results and their display are performed by a computer.

Among the commercial instruments which use this principle are the CentrifiChem (Union Carbide Corporation), Gemsaec (Electronucleonics, Inc.) and ROTOCHEM 11 (American Instrument Corporation).

Further reading: General list of analytical textbooks

CEPHALIN-CHOLESTEROL REACTION

A flocculation test for the detection of disturbances in serum proteins, which consists of adding serum to a colloidal suspension of cephalin and cholesterol. Precipitation occurs if the γ-globulin faction is raised and this precipitation is enhanced by low albumin concentrations.

See also: **flocculation tests**

CEREBROCUPREIN

A copper storage protein found in the brain

See also: **copper**

CEREBROSIDES

These are sphingolipids consisting of sphingomyelin attached to a fatty acid and a monosaccharide, usually galactose (giving galactocerebrosides), but in many cases glucose (giving glucocerebrosides).

$$\begin{array}{l} HO-CH-CH=CH\,(CH_2)_{12}\,CH_3 \\ | \\ CH-NH-CO-\text{FATTY ACID} \\ | \\ CH_2O-\text{MONOSACCHARIDE} \end{array}$$

Sulphuric acid esters of cerebrosides are known as sulphatides. In these the sulphate group is attached to the monosaccharide residue.

In the inborn error, Gaucher's disease, glucocerebrosides accumulate in various body organs due to a deficiency in the degradative enzyme, glucocerebrosidase.

See also: **Gaucher's disease**

CEREBROSPINAL FLUID (CSF)

The fluid made in the choroid plexus of the ventricles of the brain and secreted from them into the subarachnoid space around the brain and spinal cord. Specimens for laboratory examination are usually obtained from a lumbar puncture and they may be examined in a number of ways:

(1) Microbiological tests, e.g. cell count, culture, Wasserman test.

(2) Appearance. Normally CSF is a clear colourless fluid. A blood-stained fluid may indicate a subarachnoid haemorrhage or trauma during lumbar puncture. A yellow fluid (xanthochromia) may indicate bilirubin formed from haemoglobin as a result of a haemorrhage several days previously.

(3) Glucose. Low levels are found in some forms of meningitis.

(4) Protein. Changes in the composition of CSF proteins can be detected by qualitative tests, e.g. Pandy's test and the Lange test. Elevation of the CSF protein can occur in meningitis and multiple sclerosis. Specific im-

munochemical estimation of globulins can also be used for diagnosing particular conditions such as multiple sclerosis.

Further reading: Thompson, E.J., Norman, P.M. and MacDermot, J. (1975). The analysis of cerebrospinal fluid. *Br. J. Hosp. Med.*, **14**, 645

C1 ESTERASE INHIBITOR

This is an α_2-globulin which limits the action of C1, the first complement component. A genetic defect in the synthesis of C1 esterase inhibitor occurs in hereditary angioneurotic oedema. C1 esterase inhibitor can be measured immunochemically or by its ability to inhibit the hydrolysis of synthetic esters by a preparation of activated C1.

See also: **complement, hereditary angioneurotic oedema**

CHENODEOXYCHOLIC ACID

A primary bile acid formed by the liver. In the gut, bacterial action converts it to the secondary bile acid, lithocholic acid. Chenodeoxycholic acid is used therapeutically for gall stone dissolution.

See also: **bile acids and salts**

Further reading: Editorial (1978). Chenodeoxycholic acid. *Lancet*, **1**, 805

CHI-SQUARE (χ^2) TEST

A statistical test for comparing the distribution of a discrete variable in a sample with the distribution of a discrete variable in another sample. The test is for comparing actual *numbers* of occurrences rather than relative values. As an example, suppose in a drug trial, one group of patients received a new drug while another group received a placebo. All patients were examined for albuminuria and were classified as either positive or negative. The number of patients with or without albuminuria was found for each group.

The chi-square test could be used to see if there was any significant difference in the occurrence of albuminuria between the two groups. From the chi-square value, the probability of the differences between the two groups being due to chance can be found. Low probability values (<0.05, i.e. less than 1 in 20) suggest that the difference between the two groups is unlikely to be due to chance and is due to some other factor, in this case treatment with the drug.

Further reading: Swinscow, T.D.V. (1978). *Statistics at Square One*. 3rd Edn. (London: British Medical Journal)

CHLORAL HYDRATE

A hypnotic drug. It can be detected in blood and gastric aspirate by a number of screening tests.

Further reading: Meade, B.W. *et. al.* (1972). Technical Bulletin No. 24. Simple tests to detect poisons. *Ann. Clin. Biochem.*, **9**, 35

CHLORATE

Chlorates are used as weed killers and are occasionally encountered in toxicology cases, where they can cause methaemoglobinaemia and liver failure. They can be identified in gastric aspirates by their reaction with diphenylamine reagent to give a blue colour.

CHLORIDE

This is the major extracellular anion and is thus closely involved in fluid homeostasis and anion–cation balance. Its metabolism is closely linked to that of sodium.

Low serum levels

With certain exceptions (e.g. (2) below) this occurs in conditions which also cause hyponatraemia.

(1) Renal loss of chloride, e.g. renal tubular damage.

(2) Gastro-intestinal loss. Persistent vomiting results in relatively greater loss of chloride and hydrogen ions than sodium, causing hypochloraemic alkalosis.

(3) Diabetic ketosis where chloride is lost in the urine.

(4) Potassium depletion associated with alkalosis.

High serum levels

These are found in conditions such as dehydration which also cause hypernatraemia. Hyperchloraemic acidosis, associated with a low plasma bicarbonate level, is found in conditions such as renal tubular acidosis.

Sweat chloride

High sweat chloride levels are found in patients with cystic fibrosis (see **sweat test**).

Measurement of chloride in body fluids

(1) Titrimetric method (Schales and Schales). The specimen is titrated with mercuric nitrate using diphenylcarbazone as the indicator. The mercuric ions combine with chloride ions to form soluble, but virtually non-ionized mercuric chloride. When all the chloride ions have reacted with the mercuric ions, the excess Hg^{2+} ions combine with the diphenylcarbazone indicator to form a coloured complex. In this way the titration end-point is detected.

(2) Colorimetric method. Mercuric thiocyanate ($Hg(SCN)_2$) is added to the specimen, forming $HgCl_2$ and releasing thiocyanate ions which react with the Fe^{3+} of ferric nitrate solution, to produce the red coloured ferric thiocyanate. This method is used in continuous flow instruments.

(3) Coulometric method. There are several chloride meters commercially available which operate on the following principle: silver ions are generated from an electrode at a constant rate and combine with chloride ions in the specimen to form insoluble AgCl. When sufficient Ag^+ has been generated to react with all the chloride present, the additional generation of Ag^+ from the electrode results in a sudden rise in the conductivity of the solution which can be detected by a set of silver indicator electrodes.

(4) Ion-selective electrodes. These have particularly useful applications in the determination of sweat chloride in cystic fibrosis (*see* **sweat test**).

See also: **sodium**

Further reading: General list of analytical and clinical textbooks

CHLORIDE SHIFT (Hamburger shift)

CO_2 which is produced by metabolism diffuses into erythrocytes where it is converted into H_2CO_3 by carbonic anhydrase. The H_2CO_3 dissociates to give H^+ and HCO_3^-. The bicarbonate diffuses out of the cell and in order to maintain electrochemical neutrality, chloride diffuses into the red cells. Hence the term chloride shift.

CHLORPROMAZINE

A tranquilizer of the phenothiazine class of drugs. It can be detected in urine by a variety of tests. These include colour reactions (e.g. using FPN reagent), examination of absorption spectra following extraction, and various chromatographic techniques.

Futher reading: Yeoman, W.B. (1971). Toxicological analysis in the clinical chemistry laboratory. *Ann. Clin. Biochem.*, **8**, 93

Meade, B.W., *et al.* (1972). Technical Bulletin No. 24. Simple tests to detect poisons. *Ann. Clin. Biochem.*, **9**, 35

CHOLECYSTOKININ–PANCREOZYMIN

At one time there were thought to be two separate hormones, cholecystokinin which produced contraction of the gall bladder and pancreozymin which increased the secretion of pancreatic enzymes. However it is now known that a single hormone secreted by the upper small intestine performs both activities.

It can be used in conjunction with secretin to measure the capacity of the pancreas to secrete enzymes. In this test secretin is used in combination, as this stimulates the production of bicarbonate in pancreatic fluid. The secretin–cholecystokinin–pancreozymin test consists of parenteral administration of the hormone combination to the patient, followed by aspiration of the pancreatic secretions. Trypsin, amylase or lipase can be measured in the fluids as an indication of the response.

Further reading: Gowenlock, A.H. (1977). Scientific Review No. 4. Tests of exocrine pancreatic function. *Ann. Clin. Biochem.*, **14**, 61

CHOLESTEROL

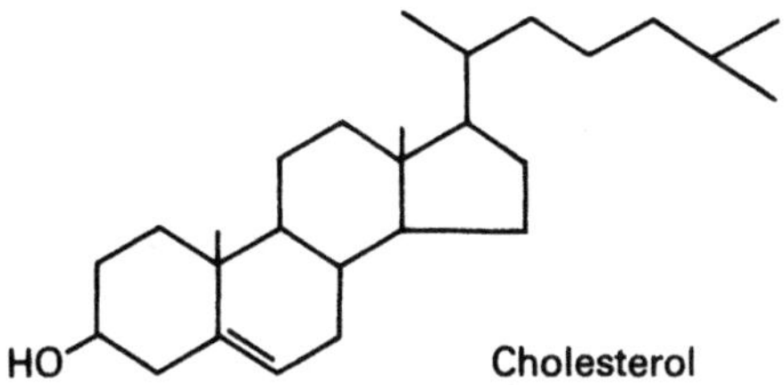
Cholesterol

High levels of cholesterol in the blood are thought to be a risk factor in the development of atherosclerotic disease. The relationship between cholesterol and high density lipoproteins may be significant in this respect (*see* **high density lipoprotein**).

The cholesterol in the body is derived from both dietary sources and endogenous synthesis.

Dietary cholesterol

Cholesterol is absorbed from the gut after its incorporation in micelles, a process for which bile salts are required. In the intestinal mucosa, the cholesterol is esterified and incorporated into chylomicrons and pre-β-lipoproteins.

Endogenously synthesized cholesterol

Cholesterol can be synthesized from acetate by most tissues, particularly the liver and small intestine. Cholesterol can be esterified and incorporated into α-and β-lipoproteins where most of the plasma cholesterol is found.

Cholesterol metabolism

Part of the body's cholesterol is used in the synthesis of bile acids and salts and a smaller proportion gives rise to steroid hormones. Some cholesterol is excreted in the bile and thus is available for reabsorption.

Increases in serum levels

Serum cholesterol levels increase with age. Among the conditions in which pathologically high levels are found are:

(1) Idiopathic hypercholesterolaemia.

(2) Biliary obstruction.

(3) Nephrotic syndrome.

(4) Hypothyroidism.

Decreases in serum levels

Decreased serum cholesterol levels are found in a great variety of conditions, e.g. liver diseases and thyrotoxicosis.

Measurement of serum cholesterol

There are two approaches to cholesterol determination.

(1) Colorimetric methods. Cholesterol reacts with strong acids to yield coloured derivatives. In most procedures acetic acid and acetic anhydride are used as solvents and dehydrating agents and sulphuric acid is used as a dehydrating and oxidizing agent. This general reaction is enhanced by various metal ions. One of the most commonly used reactions is the Liebermann–Burchard reaction in which cholesterol reacts to give a green coloured compound. In variations of this method (e.g. the Zak method) ferric ions are included in the reaction mixture and a red colour is obtained.

There can be different approaches to the pretreatment of the sample:

(a) The colorimetric reaction can be performed directly on the serum but this may give errors, due to the presence of protein and bilirubin, for which a correction may have to be made.

(b) In some techniques the lipids are extracted prior to colour development.

(c) The ratio of free to esterified cholesterol can be determined using digitonin, which selectively precipitates free cholesterol.

(2) The other basic method used in cholesterol estimation is an enzymic technique based on cholesterol oxidase. This catalyses the reaction:

$$\text{cholesterol} + O_2 \longrightarrow \text{cholest- 4-ene-3-one} + H_2O_2$$

Like the glucose oxidase reaction, the H_2O_2 formed is used to oxidize a dye, e.g. 4-aminoantipyrene, in the presence of phenol and the reaction product can be estimated colorimetrically.

Further reading: General list of analytical and clinical textbooks.

Zak, B. (1977). Review: cholesterol methodologies. *Clin. Chem.*, **23**, 1201

CHOLESTYRAMINE

An anion-exchange resin which binds bile salts and interferes with their reabsorption. In this way, it is used as a treatment for the reduction of plasma cholesterol and β-lipoprotein levels.

CHOLIC ACID

A primary bile acid formed by the liver. In the gut, bacterial action converts it to the secondary bile acid, deoxycholic acid.

See also: **bile acids and salts**

CHOLINESTERASE

Acetylcholine is a compound synthesized at nerve endings, which acts in transmitting impulses from nerve to muscle fibre. Cholinesterase is an enzyme which destroys the acetylcholine after the impulse has been transmitted, enabling further impulses to be transmitted.

In human tissues there are two enzymes with cholinesterase activity:

(1) 'True' or red cell cholinesterase (acetylcholinesterase). This is found predominantly in erythrocytes and nervous tissue.

(2) Pseudocholinesterase (cholinesterase). This is made mainly in the liver and is the enzyme which is found in serum.

Serum cholinesterase measurements are usually made in three types of clinical situations:

(1) As a test of liver function. Low levels are found in liver diseases.

(2) Low serum levels also occur in cases of poisoning by organic phosphorus compounds, e.g. certain insecticides. These inhibit both types of enzyme.

(3) In suxamethonium sensitivity. Suxamethonium is a muscle relaxant used in anaesthetics and has a similar structure to acetylcholine. Normally it is hydrolysed by cholinesterase

and this limits its action. In certain patients, however, its administration is followed by prolonged apnoea, and, in these subjects, little or no cholinesterase activity is found in the serum. Furthermore the serum enzyme in these cases is different from the normal type. This can be explained by the fact that the genes controlling the synthesis of cholinesterase can exist in at least four allelic forms, which can be differentiated on the basis of the enzyme's susceptibility to inhibition by fluoride or dibucaine (a spinal anaesthetic):

(i) The normal, most common phenotype ($E_1{}^u$). The normal enzyme is inhibited to a large extent by fluoride and dibucaine.

(ii) The atypical $E_1{}^a$ gene which results in the production of an enzyme with only weak substrate activity but more resistant to inhibition by dibucaine.

(iii) The $E_1{}^f$ gene which gives rise to a weakly active enzyme but more resistant to inhibition by fluoride.

(iv) The $E_1{}^s$ (silent) gene which results in a protein having little or no activity.

Various combinations of these genes occur but it is usually only in certain homozygotes that marked sensitivity is found. When a patient with suxamethonium sensitivity is found, their genotype can be determined by measuring their enzyme in the presence and absence of fluoride and dibucaine. Furthermore, the immediate family should be screened, to find other possible affected individuals.

Measurement of serum cholinesterase activity

The reaction catalysed by cholinesterase can be expressed as

$$\text{acetylcholine} \longrightarrow \text{choline} + \text{acetate ion} + H^+$$

(1) The H^+ can react with bicarbonate buffer, to give CO_2 which can be measured manometrically.

(2) The reaction can be followed by measuring the change in pH or by measuring the colour change of a pH indicator.

(3) The unreacted acetylcholine can be estimated colorimetrically as its hydroxamate derivative.

(4) Benzoylcholine can be used as a substrate and its decrease in absorption at 240 nm followed.

(5) Acetylthiocholine esters can be used as substrates, the thiocholine formed being measured with chromogenic disulphide reagents, e.g. Ellman's reagent.

Further reading: Beard, A.G. and Litwin, S.D. (1978). Deficiences of circulating enzymes and plasma proteins. In Stanbury, J.B., Wyngaarden, J.B. and Fredrickson, D.S. (eds.) *The Metabolic Basis of Inherited Disease*, 4th Edn., p. 1712 (New York: McGraw-Hill)

Silk, E., King, J. and Whittaker, M. (1979). Scientific review No.5. Assay of cholinesterase in clinical chemistry. *Ann. Clin. Biochem.,* **16**, 57

CHONDROITIN SULPHATE

A mucopolysaccharide excreted in the urine in some types of mucopolysaccharidoses.

See also: **mucopolysaccharides, mucopolysaccharidoses**

CHORIONIC GONADOTROPHIN

See: **human chorionic gonadotrophin**

CHROMATOGRAPHY

This is the general name given to a variety of techniques by which a group of substances are separated on the basis of their distribution between two phases, one stationary and the other moving. The varieties of chromatographic techniques may be classified in different ways:

(1) They can be classified on the basis of the physico-chemical processes involved, e.g. solvent partition, surface adsorption, molecular sieving and ion-exchange.

(2) They can be classified according to the equipment used, e.g. paper chromatography, thin-layer chromatography, column chromatography.

Thus confusion may arise in certain types of chromatography, e.g. thin layer chromatography may be either partition or adsorption chromatography depending on the solvents used.

See also: **adsorption chromatography, affinity chromatography, gas chromatography, gel filtration, high pressure liquid chromatography, ion-exchange chromatography, partition chromatography, reverse phase chromatography, thin-layer chromatography**

CHROMIUM

Chromium appears to have a role in the intermediary metabolism of carbohydrates. Deficiency of chromium may be associated with some cases of diabetes mellitus or impaired glucose tolerance.

CHROMIUM SESQUIOXIDE

A marker which can be used in faeces collection. It is determined by flame photometry or by titration against ferrous ammonium sulphate following its conversion to dichromate.

See also: **faecal markers**

CHYLOMICRONS

These are high molecular weight, lipoprotein complexes, composed mostly of triglyceride, which are synthesized in the intestinal mucosa from hydrolysed dietary fat. They are carried in the lymphatic system and pass into the bloodstream via the thoracic duct. Being high molecular weight complexes, they scatter light and are responsible for the turbidity of plasma after a fatty meal. Normally chylomicrons are cleared from the blood by the enzyme lipoprotein lipase which hydrolyses the triglycerides and releases free fatty acids which are taken up by adipose tissue. The remainder of the particle containing mostly cholesterol is further metabolized by the liver. Persistently raised serum chylomicrons are found in types I and V of the Fredrickson hyperlipoproteinaemia classification. They can be identified by their electrophoretic behaviour, since they usually remain at the origin, unlike the other lipoproteins.

See also: **hyperlipoproteinaemias, lipoproteins**

CHYLURIA

This is the presence of fat in the urine. Lipid particles (lipid emboli) can enter the blood after bone fractures and this can result in the appearance of lipid in the urine. A lymphatic disorder (filariasis) and renal damage due to stones can also cause chyluria.

Fat droplets tend to float on the surface of urine and can be visualized by staining with a fat stain, e.g. Sudan III. Quantitative estimation of triglyceride in urine can also be used.

CHYMOTRYPSIN

A proteolytic enzyme manufactured by the pancreas in the form of its zymogen, chymotrypsinogen. This is converted in the duodenum to chymotrypsin by another proteolytic enzyme, trypsin. Estimation of chymotrypsin in faeces can be of use in the diagnosis of cystic fibrosis, when enzyme secretion by the pancreas may be reduced.

See also: **cystic fibrosis**

CIRCADIAN RHYTHM (DIURNAL RHYTHM)

A regular variation within a period of 24 hours. Several substances show a circadian rhythm in their serum levels, e.g. iron and cortisol.

CITRATE

Citrate can be used as an anticoagulant because of its ability to bind calcium, which is required for the clotting process.

CITRULLINAEMIA

A rare inborn error of metabolism which may be due to a deficiency of argininosuccinic acid synthetase, the enzyme which condenses citrulline and aspartate to argininosuccinic acid. Citrulline and ammonia accumulate in the blood and CSF. Hepatomegaly and mental retardation are some of the clinical findings.

Further reading: Shih, V. (1978). Urea cycle disorders and other congenital hyperammonaemic syndromes. In Stanbury, J.B., Wyngaarden, J.B. and Fredrickson, D.S. (eds.) *The Metabolic Basis of Inherited Disease,* 4th Edn., p. 362. (New York: McGraw-Hill)

CLARK ELECTRODE

See: **oxygen**

CLEARANCE

This is the amount of blood which could theoretically be completely cleared of a substance per minute. Renal clearance can be expressed by the formula

$$\text{clearance} = \frac{U \times V}{P} \quad \text{ml/minute}$$

where U is the urine concentration of the substance, P is the plasma concentration of the substance and V is the volume of urine in millilitres passed per minute (usually calculated from a 24 hour urine collection). Since clearances are proportional to body surface area, a correction factor may have to be applied.

Estimations of the clearance of various substances can be used to assess different aspects of renal function. When a substance is excreted solely by glomerular filtration and is not reabsorbed or secreted by the tubules (e.g. inulin), estimation of its clearance, gives a true estimate of its glomerular filtration rate. Creatinine is secreted by the tubules in small amounts and therefore gives slightly higher clearance values, although these still approximate to the glomerular filtration rate. Estimation of the clearance of *p*-aminohippuric acid at low plasma levels gives an indication of renal plasma flow, whereas at higher plasma levels, estimation of its clearance gives an indication of the tubular secretory capacity.

See also: ***p*-aminohippuric acid, creatinine, glomerular filtration rate, inulin clearance test, renal plasma flow, tubular secretory capacity**

CLINISTIX

A dipstick test manufactured by the Ames company for the detection of glucose in urine. It is based on the glucose oxidase reaction, in which glucose is oxidized to gluconic acid and hydrogen peroxide. Peroxidase then catalyses the oxidation of *o*-tolidine by hydrogen peroxide to produce a blue colour.

Further reading: Kutter, D. (1977). *Rapid Clinical Diagnostic Tests*. (Munich–Vienna–Baltimore: Urban and Schwarzenberg)

CLINITEST

A tablet test, manufactured by the Ames company for the detection of reducing substances in urine. It is based on the Benedict reaction.

Further reading: Kutter, D. (1977). *Rapid Clinical Diagnostic Tests*. (Munich–Vienna–Baltimore: Urban and Schwarzenberg)

CLOFIBRATE

A drug which decreases the synthesis of cholesterol and which is therefore used in the treatment of hypercholesterolaemia.

CLOMIPHENE

A drug which stimulates the release of pituitary gonadotrophins (FSH and LH). These are normally under the control of hypothalamic releasing hormones, which in turn are under negative feedback control by the sex hormones, oestrogens and testosterone. Clomiphene is thought to act by displacing oestrogens or testosterone from their hypothalamic binding sites, thereby removing the negative feedback inhibition. The drug may therefore be used as a pituitary or hypothalamic function test. It can also be used therapeutically in the treatment of infertility to induce ovulation.

See also: **menstrual cycle**

CLOMIPHENE STIMULATION TEST

This can be used in both males and females in the investigation of suspected dysfunction of the hypothalamus or pituitary when low FSH or LH values have been found. Failure of the LH or FSH levels to rise after clomiphene administration suggests hypothalamic or pituitary dysfunction.

See also: **clomiphene**

COEFFICIENT OF VARIATION (CV)

A statistical parameter used for expressing precision. It relates the standard deviation (SD) of a group of observations to the mean as follows:

$$CV = \frac{SD}{mean} \times 100\%$$

Thus a low CV indicates good precision, a high CV poor precision.

COELIAC DISEASE

A gastro-intestinal condition in which there is sensitivity to gluten (found in wheat germ) in the diet. It causes flattening of the intestinal villi, resulting in malabsorption.

COLLAGEN

This is the major protein constituent of cartilage and other connective tissue. It is the only human protein known to contain significant amounts of hydroxyproline.

See also: **hydroxyproline**

Further reading: Editorial (1978). Collagen in health and disease. *Lancet,* **1**, 1077

COLLAGEN DISEASES (CONNECTIVE TISSUE DISEASES)

A group of diseases in which there is widespread inflammatory damage to connective tissues and blood vessels. Examples include systemic lupus erythematosus and rheumatoid arthritis. They are now known to be disorders of immunity and are also known as immune complex diseases.

See also: **rheumatoid arthritis, systemic lupus erythematosus**

Further reading: Holborow, E.J. (1978). The serology of connective tissue disorders. *Br. J. Hosp. Med.,* **19**, 250

Holborow, J. (1979). Current concepts in connective tissue disorders. *Br. J. Hosp. Med.,* **22**, 8

COLLOIDAL GOLD REACTION

A test for detecting changes in the protein composition of serum or CSF, which consists of adding a colloidal gold sol to serial dilutions of the sample and assessing the degree of precipitation of the sol.

Serum colloidal gold reaction

Positive serum colloidal gold reactions are found when there are raised γ-globulin levels with lowered albumin concentrations, e.g. hepatitis and hepatic cirrhosis.

CSF colloidal gold reaction (Lange test)

In CSF, γ-globulin is also the main precipitating agent. Other proteins tend to keep the gold in solution. Three types of abnormal precipitation patterns can be found:

(1) Paretic type found in general paralysis of the insane, in tabes, and in some cases of multiple sclerosis.

(2) Tabetic or luetic type which is found in all cases of cerebral syphilis.

(3) Meningitic type which is found in acute meningitis.

Lange colloidal gold reactions are of limited practical use. Other CSF tests give more information about the clinical condition, e.g. cell count, differential, protein, glucose and the Wassermann reaction for syphilis.

Further reading: Thompson, E.J., Norman, P.M. and MacDermot, J. (1975). The analysis of cerebrospinal fluid. *Br. J. Hosp. Med.,* **14**, 645

COLORIMETRY

The quantitative measurement of coloured solutions.

See also: **spectrophotometry**

COMPETITIVE PROTEIN BINDING (CPB)

A form of radioactive saturation analysis analagous to radioimmunoassay, the difference being that a naturally-occuring binding protein is used instead of an antibody, e.g. thyroid binding globulin for thyroxine estimation or transcortin for cortisol estimation. Apart from this difference, many of the principles of the technique are similar to radioimmunoassay.

See also: **radioimmunoassay**

COMPLEMENT

A series of plasma protein molecules, some of which have enzymic activity, which are activated by an immune complex (e.g. antibody binding to a cell). This eventually leads to a 'hole' being punched through the cell membrane, and in this way it brings about lysis and cell death. It originally received its name because it was thought to 'complement' the action of antibody.

It consists of nine major components (C1, C2, . . . C9), the components present in the largest amounts being C3 and C4. It is these two components that are measured most frequently as an indication of the overall level of complement activity.

Two complement pathways can operate, the classical pathway and the alternative pathway. The classical pathway is analagous to the coagulation cascade, activation of the C1 component eventually leading to activation of the other components. The classical pathway can be initiated by a number of substances, the most important probably being the antibody molecule. The alternative pathway can be activated by other substances. It bypasses the C1, C2 and C4 components.

Further reading: Roitt, I.M. (1977). *Essential Immunology* 3rd Edn., (Oxford, London, Edinburgh, Melbourne: Blackwell Scientific Publications)

Hoechst Pharmaceuticals, (1977). Notes on Diagnosis No. 3, *Complement*.

Whicher, J.T. (1978). Review: The value of complement assays in clinical chemistry. *Clin. Chem.*, **24**, 7

Fearon, D.T. and Austin, K.F. (1976). The human complement system: Biochemistry, biology and pathobiology. In Marks, V. and Hales, C.N. (eds.). *Essays in Medical Biochemistry* Vol. 2. p. 1. (London: The Biochemical Society and Association of Clinical Biochemists

COMPLEMENT FIXATION TEST

A test that can be used for the detection of antibodies o antigens. Antibody (or antigen) is added to a series of dilution of the patient's sample. Also present is a source of complement e.g. guinea pig serum. If antigen (or antibody) is present in th patient's sample, the resulting antigen–antibody reaction fixe the complement. An indicator system, consisting of red bloo cells coated with antibody, is then added but if no complemer remains, the cells are not lysed. On the other hand if the patient serum had no antigen (or antibody), complement would still be available and lysis of the cells would occur.

Many autoantibodies can be detected in this way, e.g. thyroid autoantibodies.

Further reading: Roitt, I.M. (1977). *Essential Immunology*. 3rd Edn. (Oxford, London, Edinburgh, Melbourne: Blackwell Scientific Publications)

CONGENITAL ADRENAL HYPERPLASIA

This is an inherited condition in which there is a deficiency of one of the enzymes of cortisol biosynthesis. The resulting low plasma cortisol levels result in high levels of ACTH production because of the absence of feedback control. This in turn results in the accumulation of androgens and cortisol precursors. The consequences of this are pseudohermaphroditism in the female child and virilization in the male child. Deficiencies of 3β-dehydrogenase, 21β-hydroxylase and 11β-hydroxylase have been described. The most common is 21β-hydroxylase deficiency, in which there is abnormal sodium loss (the salt-losing syndrome), due to impaired aldosterone production. 11β-Hydroxylase deficiency results in the excess secretion of 11-deoxycorticosterone and since this is an active mineralocorticoid, salt and water retention and hypertension result.

Biochemical diagnosis

(1) The increased production of androgens and cortisol precursors results in raised urinary 17-oxosteroids and 17-oxogenic steroids.

(2) 11-Oxygenation index. The 11-hydroxylation step is the last stage in the biosynthesis of cortisol. Any block in the pathway results in a decreased proportion of 17-oxogenic steroids with a hydroxyl group at position 11. The ratio of the steroids without an 11-hydroxyl group to those with an 11-hydroxyl group (i.e. the 11-oxygenation index) will therefore be greater than normal.

(3) In 21β-hydroxylase deficiency, the compound immediately prior to the block is 17-hydroxyprogesterone. This may be raised in the plasma where it can be measured as such. Alternatively it may be measured in the urine as its metabolite pregnanetriol.

Further reading: Bongiovanni, A.M. (1978). Congenital adrenal hyperplasia and related conditions. In Stanbury, J.B., Wyngaarden, J.B. and Fredrickson, D.S. (eds.) *The Metabolic Basis of Inherited Disease,* 4th Edn., p. 868. (New York: McGraw-Hill)

CONGENITAL ERYTHROPOIETIC PORPHYRIA

A rare type of porphyria in which the basic abnormality is confined to the erythropoietic system. Among the clinical features which can occur are photosensitivity with blistering, hirsutism and red teeth and bones. It is transmitted genetically by a recessive type of inheritance.

The condition can be diagnosed biochemically by demonstrating an excess of type I porphyrins in red cells, urine and faeces. (In other porphyrias it is type III porphyrins that are involved.)

Further reading: Meyer, U.A. and Schmid, R. (1978). The porphyrias. In Stanbury, J.B., Wyngaarden, J.B. and Fredrickson, D.S. (eds.) *The Metabolic Basis of Inherited Disease*, 4th Edn., p. 1166. (New York: McGraw-Hill)

CONGENITAL HYPERBILIRUBINAEMIA

A group of congenital disorders in which there is defective handling of bilirubin.

See also: **Crigler–Najjar syndrome, Dubin–Johnson syndrome, Gilbert's disease**

CONGO RED TEST

An *in vivo* test for the detection of amyloid in the body. Congo red dye is injected intravenously. Any amyloid material present binds the dye and its removal from the serum can be followed by taking a number of blood samples, up to an hour after injection, and measuring the dye in the serum. The test is potentially dangerous because of possible allergic reactions.

See also: **amyloid**

CONJUGATED BILIRUBIN

A term which can be used to describe the diglucuronide conjugate of bilirubin. This is water-soluble and reacts directly in the Van Den Bergh reaction for the determination of serum bilirubin. Hence the term direct bilirubin is used synonymously with conjugated bilirubin.

See also: **bilirubin**

CONN'S SYNDROME

See: **aldosteronism**

CONTINUOUS DISCRETE ANALYSIS

See: **discrete analysis**

CONTINUOUS FLOW ANALYSIS

A type of automatic analysis in which samples are mixed with reagents by means of plastic tubes, the solutions then being continuously pumped through a number of modules in which reaction takes place, finally passing to a flow-through colorimeter where the intensity of the colour is measured. The Technicon company manufactured the first commercial continuous flow analyser and since then have produced more sophisticated machines. It is these that will be described:

The Mark I AutoAnalyzer *

This consists of a number of different modules.

(1) A sampler, by which a fixed volume of the specimen is aspirated.

(2) A proportioning pump which draws reagents from the reagent bottles and mixes them with the sample. The reagent lines are segmented with bubbles by drawing air through one or more tubes. The air bubbles have a scavenging effect on the walls of the tubes and help reduce carryover between samples. When two liquids meet, a mixing coil is introduced and this inverts the liquids several times resulting in complete mixing.

(3) A dialyser in which proteins, which would interfere with many of the chemical reactions, are removed. Low molecular weight constituents diffuse into a recipient stream.

* AutoAnalyzer is a registered trade mark of Technicon Instruments Corporation, Tarrytown, New York, USA.

(4) After the dialyser, a heating bath may be used to assist in colour development.

(5) The liquid stream then passes into a continuous flow cell of a colorimeter where the bubbles are removed and the optical density is measured.

(6) The optical densities of each sample are then recorded as a series of peaks on a chart recorder. The colorimeter can be replaced by a flame photometer in the determination of sodium and potassium or a fluorimeter as for example is sometimes used in urinary oestrogen determination.

The Mark II AutoAnalyzer

This operates on the same principles as the Mark I machine but has a number of improvements.

(1) It has an improved sampler.

(2) The pump has an air bar which gives a better bubble pattern.

(3) It has a reduced dialysis time.

(4) It has an improved colorimeter with a modified flow cell.

(5) Unlike the Mark I machine, steady state conditions are achieved and these give recorder tracings as steep-sided plateaux rather than peaks. Thus Beer's law is obeyed, i.e. peak heights are proportional to concentration.

Sequential multiple analysis (SMA)

The SMA analysers, such as the SMA 6/60 or SMA 12/60 (i.e. which perform 6 or 12 tests simultaneously at 60 samples per hour), are multichannel systems making use of the improved flow characteristics and shortened dialysis path of the Mark II AutoAnalyzer. An on-line computerized system has also been produced (SMAC).

Further reading: Northam, B.E. (1970). Automatic analysis in clinical chemistry: the continuous flow system. *Br. J. Hosp. Med.*, Equipment Supplement., Nov., 20

CONWAY DISH

This is similar in appearance to a small Petri dish, the difference being that the bottom half is divided into two concentric compartments. The unit enables a gas to diffuse from liquid in one of the compartments and react with liquid in the other compartment, without the liquids themselves becoming mixed. Ammonia, carboxyhaemoglobin and ethanol are substances that can be measured in this way.

COPPER

An essential trace element for which there is a daily requirement of 2.5 mg. A normal adult contains about 100 mg. In plasma, most copper is carried by a specific copper transport protein, caeruloplasmin. Copper is also found in certain copper storage proteins; erythrocuprein (in erythrocytes), cerebrocuprein (in the brain) and hepatocuprein (in the liver).

Increases in serum levels

Increased serum copper levels are found in a variety of acute and chronic diseases, e.g. thyrotoxicosis, malignancy, biliary cirrhosis, haemochromatosis and infections. It is also raised in subjects on oestrogens or oral contraceptives.

Decreases in serum levels

Decreased serum copper levels are found in Wilson's disease and in a number of hypoproteinaemic states, e.g. malnutrition, malabsorption and nephrotic syndrome.

Measurement

(1) Atomic absorption spectrophotometry.

(2) Copper can be determined colorimetrically by its reaction with certain chromogens, e.g. biscyclohexanoneoxalyldihydrazone (Cuprizone), diethyldithiocarbamate and oxalyldihydrazide.

See also: **Wilson's disease**

Further reading: Special Issue. (1975). Trace elements in clinical chemistry. *Clin. Chem.*, **21**, No. 4

COPPER OXIDASE

Caeruloplasmin, the protein which carries copper in the blood, has oxidase activity and can therefore be referred to as copper oxidase.

See also: **caeruloplasmin**

COPROPORPHYRIN

A type of porphyrin molecule having methyl and propionate side groups (unlike uroporphyrin, which has acetate and propionate side groups, and protoporphyrin, which has methyl, vinyl and propionate side groups). Excess coproporphyrins are found in the urine and faeces in many types of porphyria and in acquired porphyrinurias (e.g. lead poisoning).

See also: **porphyrias, porphyrins**

CORRELATION COEFFICIENT (r)

This is a mathematical way of expressing the correlation between two variables, providing it is a straight line correlation. The correlation coefficient varies from –1, through 0, to +1. Complete correlation is represented by 1. If one variable increases as the other decreases, there is negative correlation; if one increases as the other increases, there is positive correlation. 0 represents complete absence of correlation.

The correlation coefficient r can be calculated from the following formula:

$$r = \frac{\Sigma (x - \bar{x})(y - \bar{y})}{\sqrt{\Sigma (x - \bar{x})^2 \, \Sigma (y - \bar{y})^2}}$$

where x and y are the two variables, and $\bar{x}$ and $\bar{y}$ are the means of the two variables.

Further reading: Swinscow, T.D.V. (1978). *Statistics at Square One*. 3rd Edn. (London: British Medical Journal)

CORTICOSTEROID-BINDING GLOBULIN

See: **transcortin**

CORTICOSTEROIDS

Steroid hormones secreted by the adrenal cortex, the principal ones being cortisol, corticosterone and aldosterone.

CORTICOSTERONE (COMPOUND B)

Along with cortisol, one of the major glucocorticoids secreted by the adrenal cortex.

CORTICOTROPHIN

See: **ACTH (adrenocorticotrophic hormone)**

CORTICOTROPHIN RELEASING FACTOR (CRF)

A hypothalamic polypeptide hormone which regulates the secretion of ACTH from the anterior pituitary. ACTH in turn stimulates the production of cortisol from the adrenal cortex. CRF secretion is controlled by three mechanisms:

(1) High levels of cortisol exert a negative feedback effect.

(2) The circadian rhythm of ACTH and cortisol due to variations in the sensitivity of the feedback and secreting systems.

(3) Physical or mental stress which can override the first two mechanisms.

See also: **adrenocorticotrophic hormone, cortisol**

CORTISOL (HYDROCORTISONE, COMPOUND F)

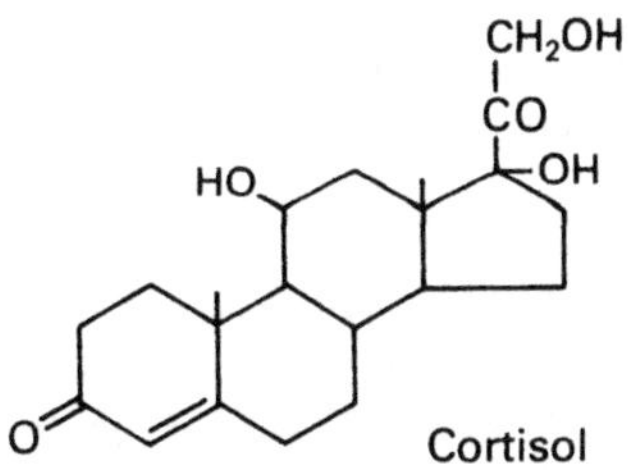

Cortisol

A steroid hormone manufactured in the adrenal cortex. Its synthesis is under the control of the pituitary hormone, ACTH. Cortisol is a glucocorticoid, stimulating gluconeogenesis, protein breakdown and lipolysis. These actions result in a raising of the blood sugar level and thus excess cortisol results in impaired glucose tolerance.

Most of the cortisol in the plasma is carried by its transport protein, transcortin and albumin. The remainder is in the free state and this is the physiologically active fraction. Cortisol levels in the plasma exhibit a circadian rhythm being lowest at about midnight and highest at about 9 a.m.

Synthesis

cholesterol ⟶ Δ^5-pregnenolone ⟶ progesterone
↓
11-deoxycortisol ⟵ 17-hydroxyprogesterone
↓
cortisol

Major breakdown products

cortisol ↗ tetrahydrocortisol, tetrahydrocortisone (50%)
cortisol ↘ cortol, cortolone (30%)

Increases in plasma levels

Increases are found in Cushing's disease.

Decreases in plasma levels

Decreases are found in Addison's disease.

Measurement

(1) By competitive protein-binding using its natural carrier protein, transcortin.

(2) By radioimmunoassay.

(3) By a flurorimetric technique. Cortisol and other corticosteroids are extracted into dichloromethane and then into a sulphuric acid–ethanol reagent, the fluorescence of which is read. This is a general method for 11-hydroxycorticosteroids.

(4) By the Porter–Silber reaction. This is the reaction of certain corticosteroids with phenylhydrazine in the presence of alcohol and sulphuric acid.

Urinary estimations of cortisol

(1) In plasma most cortisol is bound to protein and it is onl the free fraction which is physiologically active. Since onl the free fraction is filtered at the glomerulus, measuremen of urinary cortisol gives an indication of the free fraction Urinary free cortisol can be determined by competitiv

protein binding, by radioimmunoassay or by the fluorimetric determination of 11-hydroxycorticosteroids. This latter technique gives slightly higher estimates.

(2) 17-Hydroxycorticosteroids (Porter–Silber chromogens). These are measured by the phenylhydrazine–sulphuric acid reagent described above and this estimates cortisol, cortisone and 11-deoxycortisol together with their tetrahydroderivatives, i.e. cortisol precursors and their metabolites. This is used as an index of cortisol ouput.

(3) 17-Oxogenic steroids. These are urinary corticosteroids and their metabolites, measured by the Zimmermann reaction after borohydride reduction and bismuthate (or periodate) oxidation. Substances measured by this technique are similar to those measured by the Porter–Silber method with the addition of cortol and cortolone.

(4) Cortisol production rate. This consists of measuring the specific activity of a cortisol metabolite after administration of isotopically labelled cortisol. It gives similar information to that given by a 24 hour 17-oxogenic steroid determination but is subject to less interference by other chromogens.

Further reading: General list of analytical and clinical textbooks

CORTISOL PRODUCTION RATE

See: **cortisol**

CORTISONE (COMPOUND E)

A glucocorticoid secreted in small amounts by the adrenal cortex. It is not biologically active until it is converted *in vivo* to cortisol.

CORTISONE STRESSED GLUCOSE TOLERANCE TEST

This is a test for latent diabetes which can be performed if a conventional glucose tolerance test has been found to be normal. It consists of administration of cortisone to the subject prior to the glucose tolerance test. Any abnormality would suggest latent diabetes.

See also: **glucose tolerance test**

CORTOL

A major metabolite of cortisol excreted in the urine.

See also: **17-oxogenic steroids**

CORTOLONE

A major metabolite of cortisol excreted in the urine.

See also: **oxogenic steroids**

COULOMETRY

An analytical technique in which the amount of electricity passing between two electrodes in an electrochemical cell is measured. The chloride meter is an example of a coulometric technique.

See also: **chloride**

COUNTERCURRENT MECHANISM

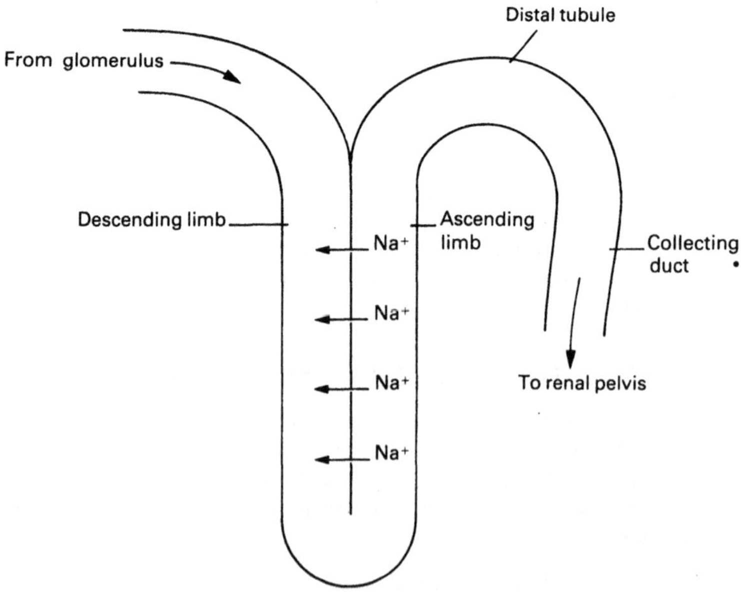

A renal mechanism for the maintainance of fluid balance. Two processes are involved:

(1) Countercurrent multiplication. This is an active process, thought to take place in the loop of Henle, in which sodium is actively pumped from the ascending limb to the descending limb while fluid is flowing through the loop. In the absence of ADH, a dilute urine is produced.

(2) Countercurrent exchange. This is a passive process which occurs in the presence of ADH. ADH renders the walls of the distal part of the tubules and the collecting ducts permeable to water. The hypo-osmolar fluid produced as a result of countercurrent multiplication is therefore concentrated.

High water intakes suppress ADH production and dilute urine is produced. Fluid deprivation leads to ADH production, rendering the walls of the distal tubules and collecting ducts permeable to water and in this way a concentrated urine is produced.

Further reading: general list of clinical textbooks

COUNTER ELECTROPHORESIS (CROSSOVER ELECTROPHORESIS)

This is similar to the double diffusion test in that antibody and antigen are placed in two adjacent holes cut out of an agar gel. They are caused to migrate towards one another by application of an electric field. Interaction between antibody and antigen results in a precipitation line. This technique thus has the advantage of being much faster than the simple double diffusion test. It is used for example for the detection of Australia antigen.

See also: **double diffusion test**

C-PEPTIDE

Insulin in the body is derived from its precursor molecule proinsulin. During the conversion of proinsulin to insulin, a small peptide (C-peptide) is released by enzymic action. Measurement of this peptide in serum provides a measure of pancreatic β-cell function, even in patients on insulin. C-peptide determination can be used in the evaluation of a number of metabolic conditions, e.g. brittle diabetes, insulinoma.

CPK

See: **creatine kinase**

C-REACTIVE PROTEIN

This is an acute phase protein found in elevated amounts in serum in many inflammatory conditions. It was so called because it reacts with the C-substance (C-polysaccharide, CPS) of pneumococcal cell walls. It has a molecular weight of 110 000–140 000, runs electrophoretically in the β or γ position depending on the conditions, and if present in sufficiently large amounts, may show as a distinct band on electrophoresis.

It is usually estimated immunochemically, e.g. radial immunodiffusion or by 'Laurell rockets'.

See also: **acute phase proteins**

CREATINE

Creatine is synthesized by the liver and, as creatine phosphate, is involved in supplying energy to a variety of cells particularly those in muscle. It is excreted as creatinine. Increased serum levels are found in myopathies and in conditions where the metabolic rate is raised, e.g. thyrotoxicosis.

Measurement

Creatinine is the anhydride of creatine. Creatine can therefore be converted to creatinine by autoclaving with strong acid.

$$\text{creatine} \xrightarrow{\text{acid}} \text{creatinine} + H_2O$$

The creatinine formed can be measured by the Jaffé reaction. Subtraction of the creatinine value of the untreated sample gives the creatine level.

See also: **creatinine**

CREATINE KINASE (CREATINE PHOSPHOKINASE, CK, CPK)

An enzyme present in heart muscle, brain and skeletal muscle. It catalyses the reaction:

$$\text{creatine} + \text{ATP} \rightleftharpoons \text{creatine phosphate} + \text{ADP}$$

Increases in serum levels

Increased serum levels are found after exercise, following a myocardial infarction (reaching a peak level about 24 hours after the event) and in muscular dystrophies.

Isoenzymes

Creatine kinase is a dimer, there being two different kinds of subunit, M or B. Thus three isoenzyme types are possible.

(1) The BB type found in brain.

(2) The MB type found in heart muscle.

(3) The MM type found in skeletal muscle.

Recently techniques have been developed which enable specific CK isoenzymes to be measured, e.g. the heart enzyme after a myocardial infarction.

Measurement of serum levels

(1) Using a coupled enzyme system as follows:

$$\text{creatine} + \text{ATP} \longrightarrow \text{creatine phosphate} + \text{ADP (creatine kinase)}$$
$$\text{phosphoenolpyruvate} + \text{ADP} \longrightarrow \text{pyruvate} + \text{ATP (pyruvate kinase)}$$
$$\text{pyruvate} + \text{NADH} \longrightarrow \text{lactate} + \text{NAD (lactate dehydrogenase)}$$

The oxidation of NADH to NAD can be followed spectrophotometrically at 340 nm. A variation of this is to measure the pyruvate formed by pyruvate kinase as its 2,4-dinitrophenylhydrazone derivative.

(2) The reaction can be measured in the reverse direction using another coupled system:

$$\text{creatine phosphate} + \text{ADP} \longrightarrow \text{creatine} + \text{ATP (creatine kinase)}$$
$$\text{ATP} + \text{glucose} \longrightarrow \text{glucose-6-phosphate} + \text{ADP (hexokinase)}$$
$$\text{glucose-6-phosphate} + \text{NADP} \longrightarrow \text{6-phosphogluconate} + \text{NADPH}$$
$$\text{(glucose-6-phosphate dehydrogenase)}$$

The formation of NADPH can be followed spectrophotometrically at 340 nm.

(3) By performing the reaction in the direction creatine phosphate $\longrightarrow$ creatine and measuring the creatine formed colorimetrically by its reaction with diacetyl and α-naphthol to form a pink coloured product.

(4) Alternatively the creatine formed can be determined fluorimetrically by its reaction with ninhydrin.

Further reading: Wilkinson, J.H. (1976). *The Principles and Practice of Diagnostic Enzymology*. (London: Edward Arnold)

CREATINE PHOSPHOKINASE

See: **creatine kinase**

CREATINE TOLERANCE TEST

A test for the investigation of myopathies in which creatine is given orally and its output in the urine is measured. Increased urine creatine excretion (i.e. reduced tolerance) is found when the muscle mass is reduced or the muscle is unable to take up creatine (e.g. hyperthyroidism and muscular dystrophy). Decreased urinary creatine excretion (i.e. increased tolerance) is found in hypothyroidism.

CREATININE

Creatinine is a nitrogenous waste product derived from creatine. It is removed from the circulation by filtration through the glomeruli, and a little is secreted by the tubules. Since creatinine production is endogenous, being dependent on muscle mass, its level in the blood is usually independent of diet, unlike urea.

Increases in serum levels

(1) Where there is increased formation of creatinine, e.g. gigantism or acromegaly.

(2) Renal failure.

Measurement of serum creatinine may not however be a sensitive indicator of early renal failure, since it may remain within normal limits even though the glomerular filtration rate is reduced by half. Also, when plasma levels are above normal, creatinine can be excreted through the tubules. The creatinine clearance test is a more useful test as it approximates to the glomerular filtration rate. This is a sensitive test for measuring renal impairment.

Measurement

Creatinine is usually estimated in body fluids by its reaction with alkaline solutions of picrate to give a red colour (Jaffé's reaction). Substances other than creatinine may react however. In urine, where the concentration of creatinine is high, this is not important, but, in the estimation of serum creatinine, the interfering chromogens may have to be removed. This may be done by isolating the creatinine on a suitable absorbent (e.g. Lloyd's reagent, Fuller's earth), leaving the interfering chromogens in solution.

Further reading: Cook, J.G.H. (1975). Technical Bulletin No. 36. Factors influencing the assay of creatinine. *Ann. Clin. Biochem.*, **12**, 219

CREATININE CLEARANCE

See: **creatinine**

CRETINISM

Cretins are children who have been hypothyroid from birth.

See: **hypothyroidism**

CRF

See: **corticotrophin releasing factor**

CRIGLER-NAJJAR SYNDROME

This is a form of congenital hyperbilirubinaemia, thought to be due to a deficiency of glucuronyl transferase, the hepatic enzyme involved in the conjugation of bilirubin.

Further reading: Schmid, R. and McDonagh, A.F. (1978). Hyperbilirubinaemia. In Stanbury, J.B., Wyngaarden, J.B., and Fredrickson, D.S. (eds.) *The Metabolic Basis of Inherited Disease,* 4th Edn., p. 1221. (New York: McGraw-Hill)

CROSSED ELECTROPHORESIS

This is the electrophoretic migration of proteins into agar gel containing antiserum. There are two types of technique.

One dimensional (electroimmunodiffusion, 'Laurell rockets')

In this technique, protein solutions are placed in a row of wells cut out of one end of agar plate into which is incorporated a specific antiserum. Electrophoresis of the sample into the gel results in antibody–antigen precipitation which takes the form of a peak or rocket. The height of the peak is related to the amount of antigen. This technique can be used to estimate the levels of specific proteins in body fluids, especially those which have a high electrophoretic mobility, e.g. the α-globulins. It can be used to measure slower moving molecules, e.g. transferrin (a β-globulin), if their mobility can be increased. This can be acheived by chemical modification of the molecule for example by carbamylation.

Two dimensional

In this technique, the protein mixture is first separated in agar by conventional electrophoresis. The plate is then turned 90° and the separated proteins are electrophoresed into agar gel containing antiserum. If the antiserum is polyvalent, a series of peaks is obtained and from these several protein antigens can be quantified simultaneously.

Further reading: Grant, G.H. and Butt, W.R. (1970). Immunochemical methods in clinical chemistry. In Bodansky, O. and Stewart, C.P. (eds.) *Advances in Clinical Chemistry* Vol. 13, p. 383. (New York: Academic Press)

Verbruggen, R. (1975). Quantitative immunoelectrophoretic methods: a literature survey. *Clin. Chem.*, **21**, 5

CROSSOVER ELECTROPHORESIS

See: **counter electrophoresis**

CRUSH SYNDROME

This occurs when large areas of muscle tissue are damaged by crushing accidents. There is severe shock with acute renal failure and uraemia. Myoglobin is found in the urine in crush syndrome.

CRYOGLOBULINS

These are proteins which precipitate or gel when cooled to below body temperature. They are usually IgG or IgM proteins, or a mixture of the two, and they are found in diseases where there is disordered immunoglobulin production. Monoclonal cryoglobulins may be found in myeloma, while polyclonal cryoglobulins are immune complexes that can be found in autoimmune diseases such as systemic lupus erythematosus or rheumatoid arthritis. Clinically, cryoglobulinaemia may present as Raynaud's syndrome with cold intolerance, peripheral gangrene and purpura.

Detection

Cryoglobulins can be detected by collecting whole blood and allowing it to clot at 37 °C. The serum is separated by centrifugation at 37 °C and then placed in a refrigerator overnight. Any cryoglobulins present will precipitate out or cause the serum to gel. Immunoelectrophoresis can help in distinguishing between monoclonal and polyclonal cryoglobulins.

Further reading: Hobbs, J.R. (1971). Immunoglobulins in clinical chemistry. In Bodansky, O. and Latner, A.L. (eds.) *Advances in Clinical Chemistry*. Vol 14, p. 220 (New York: Academic Press)

CURVE REGENERATION

This enables faster sampling rates to be attained on continuous flow analysers. The curve regenerator is connected between the colorimeter and the chart recorder of continuous flow analysers, and, by detecting the rate at which the peaks rise rather than the height of the peak as in conventional continuous flow analysis, it enables the faster sampling rates to be attained.

CUSHING'S SYNDROME

A syndrome caused by the presence of excess cortisol due to:

(1) Adenoma or carcinoma of the adrenal cortex.

(2) Bilateral adrenal hyperplasia due to

  (a) Excess ACTH secretion by the pituitary (e.g. a pituitary tumour) or

  (b) Ectopic ACTH production (e.g. from a carcinoma of the bronchus).

(3) Administration of corticosteroids.

Features

Patients with this condition present with obesity with a characteristic moon face, abdominal striae, osteoporosis, hypertension and muscular weakness. Biochemical features include impaired glucose tolerance with a diabetic type of glucose tolerance curve (because of the glucocorticoid action of cortisol) and sodium retention with potassium depletion (because of the mineralocorticoid action of cortisol).

Diagnosis

In all forms of Cushing's disease, cortisol levels in the plasma are raised and there is loss of the diurnal rhythm. This is accompanied by increased urinary excretion of cortisol. Another feature common to all forms of Cushing's disease is the failure to reduce cortisol secretion with a small dose of dexamethasone.

Finding the cause of Cushing's syndrome

(1) High levels of 17-oxogenic steroids suggests ectopic ACTH production or an adrenocortical carcinoma. Normal levels may be found in pituitary dependent Cushing's syndrome.

(2) 17-Oxosteroid excretion can be raised in adrenocortical carcinomas when androgens can be secreted in addition to corticosteroids.

(3) ACTH levels are raised in the pituitary dependent condition or in ectopic ACTH production.

(4) The metyrapone test causes a rise in urinary 17-oxogenic steroids in many cases of adrenal hyperplasia but not in the other forms of Cushing's syndrome.

Treatment of Cushing's syndrome

This varies with the aetiology of the condition, e.g. adrenalectomy in cases of adrenal carcinoma or pituitary irradiation in cases of bilateral adrenal hyperplasia.

Further reading: Hall, R., Anderson, J., Smart, G.A. and Besser, M. (1974). *Fundamentals of Clinical Endocrinology*. 2nd Edn. (London: Pitman Medical Publishing Co.)

Editorial (1977). Pituitary dependent Cushing's disease. *Br. Med. J.*, **1**, 1047

Editorial (1977). ACTH secreting lung tumours. *Br. Med. J.*, **1**, 1049

CUSUM PLOT

A form of quality control monitoring. A fixed value is assigned to a given constituent in a quality control specimen. Every time the specimen is assayed, the difference between the obtained value and the assigned value is calculated and the difference is added to the sum of previous differences (i.e. the cumulative sum or cusum). This value is plotted graphically. If the quality control is good, the cusum plot will be in the form of a horizontal line, but consistent changes in the value obtained will result in a downward or upward trend of the graph. This technique shows up small consistent disturbances which may not be detectable by inspection of a single quality control result.

See also: **quality control**

CYANIDE

Cyanide inhibits cellular respiration because of its action on respiratory enzymes and this quickly leads to death. Cyanide can be identified in the blood or urine of a suspected cyanide death by its reaction as cyanogen chloride, with pyridine and barbituric acid to give a red product.

CYANOCOBALAMIN

See: **vitamin B_{12}**

CYCLIC ADENOSINE MONOPHOSPHATE (CYCLIC AMP)

This is the substance which is thought to act as the 'second messenger' of hormone action. The 'primary messenger' is the hormone itself, e.g. ACTH, LH etc, which combines with a specific receptor on the target cell membrane, and this in turn activates adenylate·cyclase, the enzyme responsible for the synthesis of cyclic AMP. Cyclic AMP is synthesized in the cell where it modifies the cell's function.

Clinical usefulness of cyclic AMP measurements

Cyclic AMP measurements have been found to be of some use in the diagnosis of parathyroid hormone disorders. Increased urinary output of cyclic AMP is found in hyperparathyroidism, whereas a decreased output is found in hypoparathyroidism. Measurement of cyclic AMP in blood and urine after parathyroid hormone injection can also be used to differentiate between hypoparathyroidism (when there is a deficient production of parathyroid hormone) and pseudohypoparathyroidism (when parathyroid hormone is produced normally but the renal tubules do not respond to it). After parathyroid hormone administration, patients with hypoparathyroidism show an increase in urinary and plasma cyclic AMP levels but those with pseudohypoparathyroidism show a diminished response.

Measurement

Cyclic AMP can be measured by radioimmunoassay.

See also: **Ellsworth–Howard test**

CYCLOBARBITONE

A short acting barbiturate.

See also: **barbiturates**

CYSTATHIONINURIA

An inborn error of metabolism in which there are low levels of the cystathionine-splitting enzyme. This causes increased levels of cystathionine in blood and urine. Some patients have mental retardation while others appear to be normal.

Further reading: Mudd. S.H. and Levy, H.L. (1978). Disorders of transsulphuration. In Stanbury, J.B., Wyngaarden, J.B. and Fredrickson, D.S. (eds.) *The Metabolic Basis of Inherited Disease*. 4th Edn., p. 458. (New York: McGraw-Hill).

CYSTIC FIBROSIS (MUCOVISCIDOSIS, FIBROCYSTIC DISEASE OF THE PANCREAS)

An inherited disease in which pancreatic and bronchial secretions are viscid, resulting in the obstruction of these organs. It usually presents in early childhood as malabsorption with fatty stools and repeated lung infections. The sweat glands are also affected, and, because of this, the condition can be diagnosed by demonstrating an increase in the concentration of sweat sodium or chloride. In neonates, the condition may be diagnosed by the demonstration of little or no pancreatic enzymes, such as trypsin or chymotrypsin, in the faeces. The disease can be treated by giving pancreatin, an extract from the pancreas containing digestive enzymes.

Further reading: Nadler, H.L., Rao, G.J.S. and Taussig, L.M. (1978). Cystic fibrosis. In Stanbury, J.B., Wyngaarden, J.B. and Fredrickson, D.S. (eds.) *The Metabolic Basis of Inherited Disease,* 4th Edn., p. 1683. (New York: McGraw-Hill).

CYSTINE AMINOPEPTIDASE (OXYTOCINASE)

An enzyme produced by the placenta which hydrolyses oxytocin, the cystine-containing pituitary hormone which causes uterine contractions. It may therefore act in preventing the early onset of labour. Serum levels increase during pregnancy and their measurement can therefore be used as a placental function test.

Measurement

The rate of hydrolysis of a cystine-*p*-nitroanilide substrate can be used for its estimation. This involves measuring the absorbance of the liberated *p*-nitroanilide moiety by spectrophotometry.

Further reading: Wilde, C.E. and Oakey, R.E. (1975). Scientific review No. 3. Biochemical tests for the assessment of feto-placental function. *Ann. Clin. Biochem.,* **12**, 83

CYSTINOSIS

A rare inherited disorder of cystine metabolism in which there is intracellular accumulation of cystine. In the kidney this produces renal tubular damage with a non-specific aminoaciduria. Death in early childhood is the usual result.

Further reading: Schneider, J.A., Schulman, J.D. and Seegmiller, J.E. (1978). Cystinosis and the Fanconi Syndrome. In Stanbury, J.B., Wyngaarden, J.B. and Fredrickson, D.S. (eds.) *The Metabolic Basis of Inherited Disease,* 4th Edn., p. 1660. (New York: McGraw-Hill)

CYSTINURIA

An inborn error in which there is a failure of tubular reabsorption of the basic amino acids, cystine, ornithine, arginine and lysine. It is essentially a harmless inborn error, the major complication being the formation of renal stones containing cystine. Diagnosis is usually made by demonstrating the presence of excess cystine and other basic amino acids in the urine. High fluid intakes, penicillamine and alkalinization of the urine have been used in treatment of the disorder.

Further reading: Thier, S.O. and Segal, S. (1978). Cystinuria. In Stanbury, J.B., Wyngaarden, J.B. and Fredrickson, D.S. (eds.) *The Metabolic Basis of Inherited Disease,* 4th Edn., p. 1578. (New York: McGraw-Hill)

CYTOCHEMICAL HORMONE ASSAY

A recently developed form of hormone bioassay. It is based on effects produced by the hormone on sections of the target organ *in vitro,* e.g. adrenal gland if ACTH is to be measured. The hormone causes specific biochemical changes in the cells of the target organ and this can be made to result in the formation of coloured reaction products which can be quantified by means of a scanning and integrating microdensitometer.

Further reading: Rees, L.H. and Ratter, S.J. (1978). Cytochemical hormone bioassays. *Br. J. Hosp. Med.,* **19**, 229

D

DEHYDROEPIANDROSTERONE

An androgenic steroid synthesized in the ovaries, testes and adrenals. It is an intermediate in the synthesis of testosterone. Together with androstenedione, it constitutes one of the main androgens secreted by the adrenal cortex. It is one of the 17-oxosteroids which can be measured by the Zimmermann reaction.

See also: **17-oxosteroids**

DENSITOMETRY

A technique for the quantitation of coloured bands or spots on chromatograms or electrophoretic strips. Colour intensity can be measured by transmission, when light is allowed to pass through the sample, or by reflectance, when light is reflected from the coloured area. The coloured areas are represented as peaks on a chart recorder, and, from this, the area under each peak can be determined. In some instruments, the area under each peak is automatically integrated. Albumin in some laboratories is measured by densitometric techniques. Densitometers can also be used in the measurement of amniotic fluid lecithin: sphingomyelin ratios.

Further reading: General list of analytical textbooks

DEOXYCHOLIC ACID

A secondary bile acid formed by the action of intestinal bacteria on cholic acid.

See also: **bile acids and salts**

11-DEOXYCORTISOL

The immediate precursor of cortisol. It can be measured in two types of adrenal disorders:

(1) In the 11-hydroxylase deficient form of congenital adrenal hyperplasia when increased levels accumulate.

(2) As part of the metyrapone test. This drug inhibits the conversion of 11-deoxycortisol to cortisol, and is used primarily in the differential diagnosis of adrenal hypofunction. In normal subjects administration of metyrapone causes a fall in circulating cortisol. This results in the hypothalamic–pituitary axis attempting to stimulate cortisol synthesis by means of ACTH secretion. Since metyrapone has been given, the final stage of cortisol synthesis is blocked and cortisol precursors such as 11-deoxycortisol accumulate. This does not occur if there is hypothalamic–pituitary hypofunction.

See also: **congenital adrenal hyperplasia, metyrapone test**

DeRITIS RATIO

The De-Ritis ratio is the ratio of aspartate aminotransferase (AST) to alanine aminotransferase (ALT) in serum. In normal subjects this is about 1.3. AST is found in both liver cell mitochondria and cytoplasm whereas ALT is found mainly in the cytoplasm. The ratio of the two can therefore give some information on the type of liver damage. For instance in acute hepatitis the ratio is less than 1, whereas high ratios are observed in liver carcinoma.

Further reading: Schmidt, E. and Schmidt, F.W. (1976). *Brief Guide to Practical Enzyme Diagnosis,* 2nd Edn. (Mannheim: Boehringer Mannheim GmbH)

DESFERRIOXAMINE

A drug used in the treatment of haemochromatosis. It acts by chelating iron, the iron-containing chelate being subsequently excreted in the urine.

See also: **haemochromatosis**

DEXAMETHASONE SUPPRESSION TESTS

Dexamethasone is a potent cortisol analogue, which, even when administered in small amounts, is capable of suppressing ACTH secretion and thereby reducing cortisol production. It is used in the diagnosis of Cushing's syndrome, since, in all forms of this condition, administration of a small dose of dexamethasone fails to reduce cortisol secretion. In the pituitary-dependent form of the disease, this is because the feedback mechanism is insensitive and ACTH continues to be secreted. In other forms of the disease, cortisol secretion is independent of the pituitary.

The test itself can take one of two forms depending on whether urinary or plasma steroids are measured:

(1) A blood sample is taken and a small dose of dexamethasone is given. After several hours (e.g. overnight) a further blood sample is taken. Both blood samples are assayed for cortisol.

(2) The response to dexamethasone can be found by estimating the 17-oxogenic steroid content of 24 hour urines taken before and after the dose.

If no suppression is obtained, the test can be repeated with a higher dose of dexamethasone. Partial suppression may occur in pituitary-dependent Cushing's disease, due to the feedback mechanism being sensitive to the higher dose. No suppression occurs in ectopic ACTH production or adrenal carcinoma, the ACTH being already maximally suppressed.

See also: **Cushing's syndrome**

DEXTROSTIX

A reagent stick, manufactured by Ames for the estimation of blood glucose, based on the glucose oxidase reaction. The intensity of the blue colour gives an indication of the blood glucose level.

Further reading: Kutter, D. (1977). *Rapid Clinical Diagnostic Tests.* (Munich–Vienna–Baltimore: Urban and Schwarzenberg)

DIABETES INSIPIDUS

The syndrome that results from antidiuretic hormone deficiency. (A much rarer form has also been described – that of hereditary nephrogenic diabetes insipidus, where ADH secretion is normal but the renal tubules cannot respond to it.) Patients with diabetes insipidus pass large amounts of dilute urine and compensate for this loss by drinking large amounts of fluid.

Failure of the pituitary to secrete ADH may be due to pituitary or hypothalamic damage, e.g. caused by head injury or pituitary tumours.

Preparations of ADH (pitressin) are used for both diagnosis and treatment of the condition. The production of a concentrated urine, together with a reduction in its output, in response to pitressin confirms the diagnosis.

See also: **antidiuretic hormone**

DIABETES MELLITUS

A condition in which high blood glucose levels are found due to the inability of the pancreas to secrete adequate amounts of insulin. There are many forms of the disease and they have been classified into primary or secondary diabetes.

Primary diabetes

This has been divided into several degrees of severity:

(1) Potential diabetic. A person who has a normal glucose tolerance test (GTT) but has a family history of the disease.

(2) Latent diabetic. A person who has a normal GTT but at some time has had an abnormal GTT, e.g. during an infection or when pregnant.

(3) Asymptomatic diabetic. A person with a diabetic type of GTT but with no symptoms of diabetes.

(4) Clinical diabetic. A person with an abnormal GTT and with the symptoms of diabetes.

In addition to this classification, this form of the diabetes may be classed as one of two types:

(1) The juvenile-onset type. This type of diabetes is usually controlled by insulin and carries a worse prognosis.

(2) The maturity-onset type. This can be controlled by diet or oral hypoglycaemic drugs.

Secondary diabetes

This is when the diabetes is secondary to some other condition:

(1) There may be excessive amounts of hyperglycaemic agents present, e.g. glucocorticoids (as in Cushing's disease or steroid therapy) or growth hormone (as in acromegaly).

(2) Pancreatic disease, e.g. pancreatitis or carcinoma of the pancreas.

(3) There may be inhibition of insulin secretion due to excess catecholamines, e.g. stress or phaeochromocytoma.

Metabolic consequences of diabetes

(1) Hyperglycaemia and glycosuria.

(2) Because glucose has a high osmotic activity, an osmotic diuresis can occur resulting in dehydration.

(3) Insulin deficiency results in the stimulation of lipolysis, the excess free fatty acids liberated being converted to ketone bodies by the liver. This is accompanied by excess hydrogen ion production and acidosis can result.

Laboratory diagnosis of diabetes mellitus

(1) Detection of glucose in the urine.

(2) Detection of ketone bodies in the urine.

(3) Detection of high fasting and random blood glucose levels.

(4) Demonstration of an abnormal GTT.

Treatment of diabetes mellitus

(1) The juvenile-onset type of diabetes is usually treated with insulin.

(2) The maturity-onset type of diabetes can be controlled by diet or by oral hypoglycaemic drugs, e.g. the sulphonylurea or biguanide drugs.

Further reading: Notkins, A.L. (1979). The causes of diabetes. *Sci. Am.,* **241**, 56

DIAGNEX BLUE TEST

A simple test for the assessment of gastric acid secretion which dispenses with the need for collecting samples by gastric aspiration. Caffeine is first given to the patient in order to stimulate the stomach to secrete acid. Diagnex Blue granules are then given. These consist of a blue dye linked to a cation exchange resin. If hydrochloric acid has been secreted by the stomach, hydrogen ions displace the dye on the resin. The displaced dye is absorbed into the bloodstream and is eventually excreted by the kidneys. Measurement of the blue dye in the urine therefore gives an indication of the ability of the stomach to secrete hydrochloric acid.

Further reading: General list of analytical textbooks

DIASTATIC INDEX

A now somewhat dated form in which urinary amylase activity can be expressed. It is defined as the number of millilitres of 0.1% starch digested by 1 ml of urine at 37 °C in thirty minutes.

DIASTIX

A dipstick test manufactured by Ames for the semiquantitative assessment of glucose in urine. It is based on the glucose oxidase/peroxidase reactions, the hydrogen peroxide formed during the reaction oxidizing potassium iodide. The liberated free iodine blends with a background dye to give a range of colours.

Further reading: Kutter, D. (1977). *Rapid Clinical Diagnostic Tests*. (Munich–Vienna–Baltimore: Urban and Schwarzenberg)

DIBUCAINE NUMBER

Normal plasma cholinesterase (pseudocholinesterase) is 80% inhibited by the anaesthetic dibucaine, i.e. it has a dibucaine number of 80. In individuals with suxamethonium sensitivity, the cholinesterase differs from the normal form and is less susceptible to dibucaine inhibition, i.e. it has lower dibucaine numbers. This enzyme behaviour is used in phenotyping members of an affected family, since heterozygotes have dibucaine numbers intermediate between those of normals and homozygotes.

See also: **cholinesterase**

DIFFRACTION GRATING

A means by which monochromatic light can be produced in some spectrophotometers. It consists of a number of grooves cut into a metal or glass surface. When light falls on the surface, it is diffracted and spectra are obtained, the required region of which can be isolated by passing the light through a slit.

Further reading: General list of analytical textbooks

DIGOXIN

A drug which, by its actions on the heart, increases cardiac output and which is therefore used in the treatment of heart failure. It has to be used cautiously because of toxic effects. Elderly patients are particularly sensitive and overdosage occurs frequently, probably due to impaired renal function. Measurement of serum levels is therefore a useful guide to treatment. It can be measured by radioimmunoassay or enzyme-immunoassay.

Further reading: General list of analytical and clinical textbooks

DIHYDROTESTOSTERONE

An androgen formed by the reduction of testosterone in such tissues as skin, prostate and seminal vesicles. It may be a much more potent androgen than testosterone itself.

See also: **testosterone**

1,25-DIHYDROXYCHOLECALCIFEROL

The metabolically active form of vitamin D. Vitamin D is initially hydroxylated by the liver to 25-hydroxycholecalciferol and a further hydroxylation takes place in the kidney to form the dihydroxy compound. 1,25-dihydroxycholecalciferol increases calcium absorption from the intestine and, in conjunction with parathyroid hormone, releases calcium from bone.

See also: **calcium, vitamin D**

DIODRAST

A substance which can be used in a similar way to *p*-aminohippuric acid for the investigation of renal function. At low blood levels, measurement of its clearance gives an indication of renal plasma flow, whereas, at higher blood levels its clearance corresponds more to the tubular secretory capacity.

2,3-DIPHOSPHOGLYCERATE

A compound formed during glycolysis which decreases the affinity of haemoglobin for oxygen, thereby increasing oxygen delivery to tissues. This acts as a compensatory mechanism in anaemia.

Further reading: Duhm, J. (1972). The effect of 2,3-DPG and other organic phosphates on the Donnan equilibrium and the oxygen affinity of human blood. In Rorth, M. and Astrup, P. (eds.) *Oxygen Affinity of Haemoglobin and Red Cell Acid Base Status*. (Alfred Benzon Symposium, IV). p. 583. (New York: Academic Press)

DIRECT BILIRUBIN

A synonym for conjugated bilirubin, i.e. water-soluble bilirubin which reacts 'directly' in the Van den Bergh reaction without the need for an accelerator.

See also: **bilirubin**

DISACCHARIDASES AND DISACCHARIDASE DEFICIENCY

Disaccharidases are enzymes located in the intestinal mucosa which are responsible for the breakdown of disaccharides into their monosaccharide components. Among the most important are lactase (which hydrolyses lactose), maltase (which hydrolyses maltose) and sucrase (which hydrolyses sucrose). Several types of disaccharidase disorder have been described:

(1) A non-selective disaccharidase deficiency due to generalized intestinal disease.

(2) A specific lactase deficiency which may be congenital or acquired.

(3) Congenital or acquired deficiencies of sucrase-isomaltase can occur.

Patients with disaccharidase deficiency suffer from diarrhoea and abdominal discomfort due to the osmotic effects of the sugar in the intestinal tract.

Disaccharidase deficiency can be diagnosed in two ways:

(1) By measurement of the enzyme activities in a biopsied portion of intestine.

(2) By a disaccharide tolerance test. For instance if lactase deficiency is suspected, lactose is given orally. This should normally be broken down to glucose and galactose and the

former absorbed into the blood stream where it can be measured. A flat tolerance curve therefore suggests disaccharidase deficiency, although this should be followed up by a glucose tolerance test in order to check that a generalized malabsorption does not exist.

Further reading: Wilkinson, J.H. (1976). *The Principles and Practice of Diagnostic Enzymology*. (London: Edward Arnold)

Editorial. (1979). Lactase deficiency in osteoporosis. *Lancet*, **1**, 86

Editorial. (1977). Sucrose malabsorption. *Br. Med. J.*, **1**, 1558

Editorial. (1979). Lactose malabsorption and lactose intolerance. *Lancet*, **2**, 831

Gray, G.M. (1978). Intestinal disaccharidase deficiencies and glucose-galactose malabsorption. In Stanbury, J.B., Wyngaarden, J.B. and Fredrickson, D.S. (eds.) *The Metabolic Basis of Inherited Disease*. 4th Edn., p. 1526 (New York: McGraw-Hill)

DISC ELECTROPHORESIS

A high resolution electrophoretic technique in which the electrophoretic matrix, usually polyacrylamide, is contained in a cylindrically shaped cell. The electrophoretic matrix contains regions of gel of different pore size. This discontinuity and the fact that the separated components have a disc shape give the technique its name.

Further reading: General list of analytical textbooks

DISCONTINUOUS DISCRETE ANALYSIS

See: **discrete analysis**

DISCRETE ANALYSIS

A type of automatic analysis in which samples are processed in separate reaction tubes. A typical discrete analysis system consists of a series of reaction tubes into which the sample is dispensed together with reagent. Incubation or addition of further reagents can follow and then the optical density can be measured by direct reading or by drawing the contents of the tube into a flow-cell colorimeter. At the end of the analysis, the tubes can be washed automatically or discarded.

Discrete analysis itself can be divided into 'discontinuous discrete' analysis, where the operator moves the tubes from one stage of the process to the next, and 'continuous discrete' analysis, where this occurs automatically.

One of the major problems of discrete analysis is deproteinization. This can be overcome by choosing methods which do not require the removal of proteins. If protein removal is unavoidable a centrifugation step is included in the process.

Examples of discrete analysers are the Vicker's M300, the Abbott Bichromatic Analyzer and the Hycel Mark X.

Further reading: Northam, B.E. (1971). Automatic analysis in clinical chemistry; discrete analysis systems. *Br. J. Hosp. Med.,* Equipment Supplement May, 44

DIURNAL RHYTHM

See: **circadian rhythm**

DOPAMINE (DIHYDROXYPHENYLETHYLAMINE)

A catecholamine which is an intermediate in the synthesis of adrenaline and noradrenaline. Excessive secretion of dopamine and its metabolites can occur in neuroblastoma.

See also: **catecholamines**

DORIDEN

See: **glutethimide**

DOUBLE DIFFUSION TEST (OUCHTERLONY TECHNIQUE)

A test for detecting the presence of antibodies or antigens in biological fluids. Antibody and antigen solutions are placed in separate wells cut out of an agar plate. Diffusion of the antibody and antigen towards each other occurs and, if there is interaction, the immune complexes precipitate in the form of a line. Thus for instance if C-reactive protein was suspected in a sample of serum, the serum would be placed in the one well and C-reactive protein antiserum in the other. On the other hand if antibodies were suspected in a serum sample, e.g. thyroid autoantibodies, the sample would be placed in one well and antigen (in this case thyroid extract) would be placed in the other.

Further reading: Grant, G.H. and Butt, W.R. (1970). Immunochemical methods in clinical chemistry. In Bodansky, O. and Stewart, C.P., (eds.) *Advances in Clinical Chemistry*. Vol. 13, p. 383. (New York: Academic Press Inc.)

DUBIN–JOHNSON SYNDROME

A form of congenital hyperbilirubinaemia which is thought to be due to defective excretion of conjugated bilirubin. The condition is relatively harmless.

Further reading: Schmid, R. and McDonagh, A.F. (1978). Hyperbilirubinaemia. In Stanbury, J.B., Wyngaarden, J.B. and Fredrickson, D.S. (eds) *The Metabolic Basis of Inherited Disease*. 4th Edn., p. 1221. (New York: McGraw-Hill)

DUCHENNE MUSCULAR DYSTROPHY

A form of muscular dystrophy in which very high serum levels of creatine phosphokinase are found. It is a genetically determined disease and clinically normal carriers of the condition can also be detected by raised serum CPK levels.

Further reading: Appel, S.H. and Roses, A.D. (1978). The muscular dystrophies. In Stanbury, J.B., Wyngaarden, J.B. and Fredrickson, D.S. (eds.) *The Metabolic Basis of Inherited Disease,* 4th Edn., p. 1260. (New York: McGraw-Hill)

DUMPING SYNDROME

A consequence of gastrectomy when patients feel faint and develop abdominal discomfort after a meal. The rapid passage of food in fluid form into the duodenum, is thought to be responsible for these effects. This fluid has a high osmotic content and this results in water passing from body fluids into the intestinal lumen, causing the abdominal symptoms. The resulting reduction in plasma volume causes faintness.

Further reading: General list of clinical text books

DUODENAL INTUBATION

The passage of a tube, via the mouth, into the duodenum in order to collect duodenal fluid which contains pancreatic secretions. This procedure is used in pancreatic function tests.

See also: **Lundh test, secretin-cholecystokinin-pancreozymin stimulation test, secretin stimulation test**

E

ECTOPIC HORMONE PRODUCTION

The synthesis of a hormone at a site where it is not normally produced. Various tumours, particularly carcinoma of the bronchus, can result in the secretion of hormonal substances. Among the hormones which can be produced by tumours are antidiuretic hormone, parathyroid hormone, adrenocorticotrophic hormone, gastrin (in the Zollinger–Ellison syndrome) and erythropoietin.

See also: **antidiuretic hormone, Cushing's syndrome, gastrin, hyperparathyroidism, Zollinger–Ellison syndrome**

Further reading: Hall, R., Anderson, J., Smart, G.A. and Besser, M. (1974). *Fundamentals of Clinical Endocrinology*. 2nd Edn. (London: Pitman Medical Publishing Co.)

Editorial. (1977). ACTH secreting lung tumours. *Br. Med. J.*, **1**, 1047

EFFECTIVE THYROXINE RATIO*

A means of expressing serum thyroxine values which takes into consideration any abnormalities in protein binding. It is thus analogous to the free thyroxine index.

See also: **thyroxine**

EHRLICH'S REAGENT

This consists of *p*-dimethylaminobenzaldehyde in hydrochloric acid. It is used in the estimation of urinary urobilinogen (which gives a red colour), porphobilinogen, (giving a red colour which can be extracted into organic solvents) and indican (when the red chromogen can be extracted into alkali).

* Registered trademark of Mallinckrodt Chemical Works, St. Louis, Missouri, USA.

ELECTROENDOSMOSIS (ENDOSMOSIS)

A phenomenon that can occur during electrophoresis, in which the molecules to be resolved do not migrate, or perhaps even move in the opposite direction to that which is expected. It is due to the support medium (e.g. agar or cellulose acetate) taking on a net negative charge because of the adsorption of hydroxyl ions. This leaves a net positive charge in the form of hydroxonium ions (H_3O^+) in the solvent. These ions migrate towards the cathode and this results in the net movement of water in this direction, carrying with it some of the more weakly charged molecules.

ELECTROIMMUNODIFFUSION

See: **crossed electrophoresis**

ELECTROLYTES

A term for the anions and cations of the body fluids, although, in practice, the term is usually confined to the commonly measured electrolytes, sodium, potassium, bicarbonate and chloride.

ELECTROPHORESIS

The separation of a mixture of substances by their different rates of migration in a charged field.

See: **agarose, cellulose acetate, disc electrophoresis, isoelectric focusing, isotachophoresis, starch gel electrophoresis, zone electrophoresis**

Electrophoresis can also be used in conjunction with immunochemical techniques for the identification and quantitation of proteins.

See: **crossed electrophoresis, counter electrophoresis, immunoelectrophoresis**

ELISA

See: **enzyme-immunoassay**

Measurement

(1) Microdiffusion methods in which ethanol is released from the sample and absorbed by an acid dichromate solution. Ethanol is oxidized to acetic acid and at the same time the yellow dichromate is reduced to green chromic ion. This technique can be carried out in a Conway dish.

(2) Ethanol can be measured enzymically by its oxidation by alcohol dehydrogenase to acetaldehyde. At the same time, NAD is reduced to NADH and this can be followed spectrophotometrically.

(3) Gas–liquid chromatographic methods.

Further reading: General list of analytical textbooks

ETHYLENEDIAMINE TETRACETIC ACID (EDTA, VERSENE, SEQUESTRENE)

A compound which can be used as an anticoagulant by virtue of its ability to chelate calcium, which is essential for the clotting mechanism. In most cases it is the dipotassium salt which is used for this purpose. EDTA is also used in the treatment of lead poisoning because of its ability to chelate this metal.

EUGLOBULIN

An outmoded term for part of the globulin fraction. It refers to those proteins that are precipitated between 28% and 33% ammonium sulphate fractionation. (Globulins precipitating between 33% and 50% ammonium sulphate fractionation are referred to as pseudoglobulins.)

The major components of the euglobulin fraction are IgM, some IgG, low density lipoprotein, caeruloplasmin and β_1c-globulin.

See also: **proteins**

EVANS BLUE TEST

A test for the measurement of plasma volume which consists of intravenous injection of a known amount of the dye Evans Blue, followed by measurement of its subsequent dilution in the blood.

EWALD MEAL

A gastric function test which consists of measuring acid output after stimulation with toast and water (or tea). It is now rarely used.

EXTINCTION

See: **absorbance**

EXTRACELLULAR FLUID

One of the two main fluid compartments into which the body can be theoretically divided (the other being the intracellular fluid). Extracellular fluid can be subdivided into the intravascular fluid (i.e. plasma) and interstitial fluid (i.e. the fluid between the tissue cells). Plasma is separated from interstitial fluid by the capillary wall which acts as a semipermeable membrane, allowing the passage of water and small molecules, but not the larger molecules such as proteins.

EXTRINSIC FACTOR

See: **vitamin B_{12}**

F

F TEST

A statistical test which can be used for the comparison of precisions. It is used, for example, to determine whether a significant difference in precision exists between two different methods.

Further reading: Moroney, M.J. (1953). *Facts From Figures.* 2nd Edn. (Harmondsworth: Penguin Books)

FABRY'S DISEASE

An inborn error of glycosphingolipid catabolism characterized by the accumulation of a ceramide derivative in blood and many tissues. It is due to a deficiency of one of the catabolic enzymes, α-D-galactoside galactohydrolase. It is transmitted by an X-linked gene.

Further reading: Desnick, R.J., Klionsky, B. and Sweely, C.C. (1978). Fabry's disease (α-galactosidase A deficiency). In Stanbury, J.B., Wyngaarden, J.B. and Fredrickson, D.S. (eds.) *The Metabolic Basis of Inherited Disease.* 4th Edn. p. 810. (New York: McGraw-Hill)

FAECAL FAT

Normally, faeces contain only small amounts of fat, this being mainly derived from endogenous sources such as desquamation or intestinal secretions. High faecal fat levels (steatorrhoea) are found in certain types of malabsorption, e.g. coeliac disease, pancreatic disease and idiopathic steatorrhoea.

Laboratory investigation of faecal fat

(1) A simple screening test for fat globules consists of mixing a portion of the faeces with a fat stain, e.g. Sudan III and examining it under a microscope. Fat globules stain orange or red with Sudan III.

(2) A quantitative faecal fat estimation may be performed. Faeces collections are made over three or five days, markers such as carmine or chromium sesquioxide being sometimes used to time the beginning and end of the collection. The patient may be placed on a known fat diet during the collection, enabling the fat balance to be determined.
In one of the most common methods for the estimation of faecal fat, the triglycerides are hydrolysed with potassium hydroxide. The liberated free fatty acids are extracted into petroleum and estimated by titration against a known concentration of alkali.

See also: **triolein-oleic acid absorption test**

FAECAL MARKERS

Substances which, when taken orally, can be used for the timing of faecal collections or for correcting day to day variations in the amount of faeces passed. For instance if the dye carmine is given over several days, all the specimens containing the dye, can be taken to represent the faecal collection. Chromium sesquioxide is another commonly used marker.

FANCONI SYNDROME

A generalized aminoaciduria that results from proximal renal tubular damage. Failure to reabsorb phosphate and glucose also occurs. The syndrome can be caused by endogenous poisons (e.g. copper in Wilson's disease) or exogenous poisons (e.g. cadmium) damaging the renal tubule.

See also: **aminoaciduria**

FAST HAEMOGLOBINS

See: **glycosylated haemoglobins**

FAT ABSORPTION TEST

A test of fat malabsorption which consists of giving a large oral amount of butter to the patient. In normal subjects absorption of the butter is followed by a significant rise in the chylomicron level in the serum, which can be measured quantitatively, for example by nephelometry. Failure of the serum chylomicron level to rise significantly is therefore suggestive of fat malabsorption.

See also: **triolein-oleic acid absorption test**

FAVISM

A type of glucose-6-phosphate dehydrogenase deficiency in which the red blood cells of a susceptible person haemolyse after the person has been in contact with fava beans (from the bean plant, *Vicia fava*). This disorder is mainly found in Southern Europe.

Further reading: Beutler, E. (1978). Glucose-6-phosphate dehydrogenase deficiency. In Stanbury, J.B., Wyngaarden, J.B. and Fredrickson, D.S. (eds.) *The Metabolic Basis of Inherited Disease*. 4th Edn., p. 1430. (New York: McGraw-Hill)

FEARON'S TEST

A test for the detection of reducing disaccharides (e.g. lactose) in urine. The test is based on the reaction of these sugars with methylamine to produce a red colour.

Further reading: General list of analytical textbooks

FERRITIN

A storage form of iron consisting of ferric hydroxide and ferric phosphate attached to the protein apoferritin. It is found in the reticulo-endothelial system, the liver and muscle. The ferritin in plasma is in equilibrium with the ferritin in iron stores. Thus low plasma levels may suggest iron deficiency while high plasma levels suggest iron overload. It can be measured by immunoradiometric assay.

See also: **iron**

Further reading: Editorial (1979). Serum ferritin. *Lancet*, **1**, 533

FETOPLACENTAL FUNCTION TESTS

See: **pregnancy**

α_1-FETOPROTEIN

An α_1-globulin normally only found in detectable quantities in fetal serum. It is measured in two clinical situations:

(1) In the detection of fetuses with open spina bifida or anencephaly when α-fetoprotein leaks into the amniotic fluid in increased amounts. Increased maternal serum levels may also be found in these cases and this forms the basis of screening programmes for the detection of affected fetuses with a view to abortion.

(2) In the diagnosis of patients with certain carcinomas, especially primary carcinoma of the liver and teratomas.

Measurement

α-fetoprotein is usually measured by radioimmunoassay.

Further reading: Kohn, J. and Weaver, P.C. (1974). Serum alpha$_1$-fetoprotein in hepatocellular carcinoma. *Lancet,* **2**, 334

Kohn, J., Orr, A.H., McElwain, T.J., Bentall, M. and Peckham, M.J. (1976). Serum alpha$_1$-fetoprotein in patients with testicular tumours. *Lancet,* **2**, 433

Editorial. (1976). Antenatal diagnosis of spina bifida. *Br. Med. J.,* **1**, 414

Editorial. (1979). Origin of material serum α-fetoprotein. *Lancet,* **2**, 999

FIBRINOGEN

A plasma protein, synthesized by the liver. It is converted to fibrin during the coagulation process to form the matrix of the blood clot. Thrombin is required for this conversion. Increased plasma fibrinogen levels are found after trauma and in many inflammatory diseases. The increases in the plasma fibrinogen level account, at least in part, for the increased erythrocyte sedimentation rate in such cases. Decreased plasma levels are found in liver diseases and in the genetic defect congenital hypofibrinogenaemia.

Measurement

(1) Measuring the clotting time in the presence of thrombin.

(2) By selectively precipitating fibrinogen and measuring its protein content, for example by the biuret reaction.

(3) By adding sufficient salt solution (e.g. sodium sulphite) to cause precipitation of the fibrinogen without precipitating the other plasma proteins. The turbidity of the suspension is then measured.

Further reading: General list of analytical and clinical textbooks

FIBROCYSTIC DISEASE

See: **cystic fibrosis**

FIGLU TEST

See: **formiminoglutamic acid test**

FLAME PHOTOMETRY

A term used to describe the estimation of certain metals by either atomic absorption spectrophotometry (*q.v.*) or atomic emission spectrophotometry (*q.v.*).

FLOCCULATION TESTS

Tests in which abnormalities in the protein concentrations in serum or CSF cause turbidity or flocculation of certain reagents, e.g. colloidal gold, thymol, zinc sulphate and cephalin-cholesterol. See separate entries for these tests.

FLUORIDE

Fluoride is sometimes used in the treatment of osteoporosis and Paget's disease of the bone. It may also be encountered in cases of accidental or deliberate poisoning following ingestion of fluoride-containing substances. It can be measured in biological fluids by its colorimetric reaction with alizarin complexone and either cerium or lanthanum. Alternatively a fluoride ion selective electrode may be used.

FLUORIDE NUMBER

Normal cholinesterase is inhibited to a large extent by fluoride (i.e. it has a high fluoride number). In individuals with suxamethonium sensitivity, the cholinesterase differs from the normal form and it is less susceptible to inhibition (i.e. has lower fluoride numbers). This fact can be used in phenotyping members of an affected family since heterozygotes have fluoride numbers intermediate between those of normals and homozygotes.

See also: **cholinesterase**

FLUORIDE-OXALATE

A combination of preservative and anticoagulant which is used in the collection of blood for certain estimations particularly glucose. Fluoride inhibits glycolysis while oxalate acts as an anticoagulant by complexing with calcium, which is required for the coagulation process.

FLUORIMETRY

The quantitative measurement of fluorescence. A molecule fluoresces when it is excited by light at one wavelength, and, in returning to a low energy state, emits light of a longer wavelength (i.e. lower energy). Mercury and xenon lamps can provide the high energy incident light and monochromators can be used to isolate a specific excitation wavelength. The fluorescence is usually measured at right angles to the light beam in order that transmitted light does not interfere. A second monochromator can be used to isolate fluorescent light of a specific wavelength.

One of the major advantages of fluorimetric techniques is their extreme sensitivity. They are used in the estimation of a variety of substances, including cortisol, oestrogens and phenylalanine.

See also: **fluoroimmunoassay**

Further reading: Rubin, M. (1970). Fluorimetry and phosphorimetry in clinical chemistry. In Bodansky, O. and Stewart, C.P. (eds.) *Advances in Clinical Chemistry,* Vol. 13, p. 163 (New York: Academic Press)

FLUOROIMMUNOASSAY

A technique for the measurement of various substances based on the labelling of materials with fluorescing compounds, e.g. fluorescein. The techniques are analogous to radioimmunoassay, the difference being that the antigen is labelled with a fluorescent material instead of a radioactive isotope. Some fluoroimmunoassay techniques do not depend on the separation of the free and bound antigen. For instance in the fluorescence-quenching type of methods, a decrease in fluorescence is measured as the labelled antigen combines with the antibody. However, in fluorescence-enhancement techniques, it is an increase in fluorescence which is measured as the labelled antigen combines with the antibody.

Fluoroimmunoassays have been used to measure a variety of compounds including the C3 component of complement, thyroxine and gentamicin. One of the main advantages of fluoroimmunoassay over radioimmunoassay is that it dispenses with the need for potentially hazardous radioactive materials. One of the major disadvantages of the technique is decreased sensitivity due largely to background fluorescence, e.g. from serum.

See also: **immunofluorescence**

Further reading: Soini, E. and Hemmila, I. (1979). Fluoroimmunoassay. Present status and key problems. *Clin. Chem.*, **25**, 353

FOLIC ACID

A vitamin of the B group found in some vegetables and meats. It is involved in purine and pyrimidine synthesis (constituents of DNA and RNA) where it has a role in the transfer of one of the carbon units. Folate participates in many of these reactions in the form of its reduced derivative, tetrahydrofolic acid. Deficiency of the vitamin, either through malabsorption or a dietary deficiency, causes megaloblastic anaemia. Certain drugs, e.g. phenytoin also interfere with folate metabolism. Folic acid can be measured by competitive protein binding or by bioassay, using folic acid-requiring strains of micro-organisms. Folate deficiency can also be investigated *in vivo* by the FIGLU test.

Further reading: Chanarin, I. (1977). Folates, cobalamins and their interrelationship in man. In Marks, V. and Hales, C.N. (Eds.) *Essays in Medical Biochemistry*. Vol. 3, p.1. (London: The Biochemical Society and the Association of Clinical Biochemists)

FOLIN-CIOCALTEU REAGENT

A complex phosphotungstomolybdic acid reagent which oxidizes phenolic compounds under alkaline conditions and is itself reduced from a yellow to a blue colour. Examples of its use are the estimation of total protein by its reaction with tyrosine residues and the measurement of phenol liberated by alkaline phophatase in the King–Armstrong method.

See also: **alkaline phosphatase**

FOLIN-WU METHOD

A method for the estimation of glucose in body fluids. In the technique a protein-free filtrate of the sample is prepared using tungstic acid. Glucose in the filtrate reduces cupric ion to cuprous ion. The cuprous ion then reduces phosphomolybdic acid to molybdenum blue which can be estimated colorimetrically.

See also: **glucose**

FOLLICLE-STIMULATING HORMONE (FSH)

A gonadotrophin secreted by the anterior pituitary in response to a specific releasing factor from the hypothalamus (LH/FSH releasing hormone). FSH has a glycoprotein structure. In the female, FSH promotes growth and maturation of the ovarian follicle, while, in males, it stimulates spermatogenesis.

Plasma FSH levels reach a peak in the follicular phase of the menstrual cycle and then decline towards midcycle. At ovulation a sharp peak occurs followed by a further drop to basal levels. Oestradiol is involved in a feedback mechanism governing FSH release.

High serum FSH levels suggest ovarian failure in females or primary testicular failure in males. Low FSH levels suggest hypothalamic or pituitary dysfunction.

FSH is usually measured by radioimmunoassay.

Further reading: Hall, R., Anderson, J., Smart, G.A. and Besser, M. (1974). *Fundamentals of Clinical Endocrinology*, 2nd Edn. (London: Pitman Medical Publishing Co.)

FORMALDEHYDE STABLE ACID PHOSPHATASE

A term which is synonymous with prostatic acid phosphatase. The total acid phosphatase activity in serum is derived from a number of different sites, e.g. the red blood cell, bone and prostate. The prostatic isoenzyme, unlike the red cell enzyme, is stable in the presence of formaldehyde. This property is used to measure the prostatic isoenzyme.

See also: **acid phosphatase**

FORMIMINOGLUTAMIC ACID (FIGLU) TEST

The normal metabolism of histidine contains a step in which formiminoglutamic acid is converted to glutamate by an enzyme which uses folate as a cofactor. In individuals with folate deficiency, administration of oral histidine results in a greater than normal urinary excretion of FIGLU. The FIGLU is usually determined enzymically or by electrophoresis.

See also: **folic acid**

FOUCHET'S TEST

A test for the detection of bilirubin in urine. It is based on the oxidation of bilirubin, by ferric chloride, to produce biliverdin, which can be seen as a green coloration. Various drugs, e.g. chlorpromazine, can interfere by producing a purple colour.

FRANKLIN'S DISEASE

See: **heavy chain diseases**

FREDRICKSON CLASSIFICATION

A classification of the hyperlipoproteinaemias based on the electrophoretic behaviour of the plasma lipoproteins.

See also: **hyperlipoproteinaemias**

FREE CORTISOL

The majority of the plasma cortisol is protein bound, only a small proportion being in the free state. This free cortisol is filtered at the glomerulus and passes into the urine where it can be measured, for example by radioimmunoassay. Urinary free cortisol determinations therefore correlate with the plasma free cortisol and the cortisol secretion rate. Increased urinary free cortisol levels are found in cases of Cushing's syndrome.

See also: **cortisol, cortisol production rate**

FREE FATTY ACIDS

See: **non-esterified fatty acids**

FREE THYROXINE INDEX (FTI)

Although most circulating thyroxine is protein bound, it is the free fraction which is physiologically active. The free thyroxine index corrects the total serum thyroxine for any abnormalities there may be in protein binding (e.g. the low TBG levels found in hypoproteinaemic states or the high TBG levels found in pregnancy). It can be calculated from the total serum thyroxine and the T_3 resin uptake test value.

See also: **thyroxine, T_3 uptake test**

FREUND'S ADJUVANT

An emulsion of mineral oil and lyophilized bacteria, which, when injected into animals, stimulates antibody production. When used in conjunction with a particular antigen, it is capable of stimulating the production of an antiserum of a higher titre than that which would be obtained by injecting the antigen alone.

FRUCTOSE (LAEVULOSE)

A monosaccharide obtained from dietary sources, either as the free sugar or as a part of the sucrose molecule. It is normally metabolized by the liver and kidney to glucose. Two inborn errors of fructose metabolism have been described:

(1) Essential fructosuria which is a benign condition caused by a deficiency of the enzyme fructokinase.

(2) Hereditary fructose intolerance which is a more serious condition in which there are metabolic disturbances, e.g. hypoglycaemia, after fructose ingestion. It is due to a deficiency of the enzyme fructose-1-phosphate aldolase.

Fructosuria occurs in both these conditions and this can be detected by Seliwanoff's test.

See also: **essential fructosuria, hereditary fructose intolerance**

FRUCTOSE TOLERANCE TEST

A test for the detection of glycogen storage disease type I (Von Gierke's disease). Infusion of fructose in normal subjects is followed by a rise in blood glucose as a result of the conversion of fructose to glucose in the liver. This does not occur in glycogen storage disease type I where there is a deficiency of glucose-6-phosphatase, the enzyme which converts glucose-6-phosphate (formed from fructose) to glucose.

FSH

See: **follicle stimulating hormone**

G

GALACTOKINASE

An enzyme involved in the conversion of galactose to glucose. A deficiency of the enzyme occurs in one of the two forms of galactosaemia.

The enzyme can be measured in red blood cells by following the conversion of $[^{14}C]$galactose to $[^{14}C]$galactose-1-phosphate.

See also: **galactosaemia**

Further reading: Wilkinson, J.H. (1976). *The Principles and Practice of Diagnostic Enzymology*. (London: Edward Arnold)

GALACTOSAEMIA

An inborn error in which there is a defect in the metabolism of galactose. Galactose is normally metabolized to glucose via the following pathway.

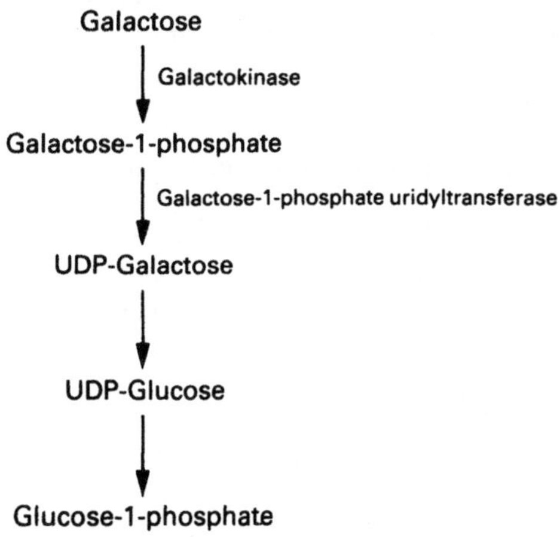

One form of galactosaemia is due to a deficiency of galactokinase but the most common form is due to a deficiency of galactose-1-phosphate uridyltransferase (hexose-1-phosphate uridylyltransferase). The condition presents in infancy with symptoms that include vomiting, diarrhoea, failure to thrive, mental retardation, renal tubular damage and hepatomegaly leading to jaundice.

Biochemical diagnosis

(1) By the demonstration of galactose in the urine. Galactose is a reducing sugar and therefore gives positive reactions with Benedict's reagent or Clinitest tablets. It can be distinguished from glucose by the fact that only glucose will give a positive reaction with Clinistix (a reagent stick for the specific detection of glucose, based on the glucose oxidase reaction). The presence of galactose can be confirmed by chromatography.

(2) By demonstrating a deficiency of one of the two enzymes in red cells.

Treatment

This involves a dietary limitation of milk and milk derived products (which contain lactose).

Further reading: Segal, S. (1978). Disorders of galactose metabolism. In Stanbury, J.B., Wyngaarden, J.B. and Fredrickson, D.S. (eds.) *The Metabolic Basis of Inherited Disease*. 4th Edn., p. 160. (New York: McGraw-Hill)

GALACTOSE

This monosaccharide can be both synthesized by the body and derived from dietary sources (especially from the disaccharide, lactose, where it constitutes one of the monosaccharide components). It is metabolized by the liver to glucose and glycogen. A failure to metabolize galactose occurs in the inborn error of metabolism, galactosaemia.

Measurement

Galactose can be measured in body fluids, using the enzyme galactose oxidase in a reaction analogous to the glucose oxidase reaction for the determination of glucose.

See also: **galactosaemia**

GALACTOSE-1-PHOSPHATE URIDYLTRANSFERASE

An enzyme involved in the conversion of galactose to glucose. It is the enzyme which is deficient in the most common form of galactosaemia.

Measurement

The enzyme can be measured in red blood cells by one of several techniques.

(1) One of the most extensively used methods is based on the measurement of UDP-glucose before and after incubation of the red cell haemolysate with galactose-1-phosphate. UDP-glucose is measured by the enzyme UDP-glucose dehydrogenase which converts UDP-glucose to UDP-glucuronic acid, using NAD as a cofactor. This reaction can be followed spectrophotometrically at 340 nm.

(2) The enzyme can also be estimated by measuring the respiration of red cell haemolysates using galactose-1-phosphate as the substrate. This involves measuring oxygen uptake in capillary tubes.

See also: **galactosaemia**

Further reading: Wilkinson, J.H. (1976). *The Principles and Practice of Diagnostic Enzymology*. (London: Edward Arnold)

GALACTOSE TOLERANCE TEST

Since galactose is mainly metabolized by the liver, the body's handling of an administered oral load of galactose can be used as a test for liver damage. In such conditions the blood galactose level reaches much higher values than in normal subjects.

GALL STONES

Concretions formed in the gall bladder. They can be composed of cholesterol, bile pigments or a mixture of the two. Many factors have been implicated in the formation of gall stones including infections, disturbances in the ratio of bile acids to cholesterol, stagnation of bile, and pH changes.

GANGLIOSIDES

These are glycosphingolipids which consist of sphingosine, a fatty acid and an oligosaccharide containing sialic acid. They are formed from ceramides by the attachment of sialic acid residues. Accumulation of gangliosides is found in two inborn errors of lipid storage, Tay–Sachs disease and generalized gangliosidosis.

GANGLIOSIDOSES

Inborn errors of lipid storage characterized by the accumulation of various types of gangliosides.

See: **generalized gangliosidosis, Tay–Sachs disease**

GAS CHROMATOGRAPHY

A form of chromatography in which molecules can be resolved on the basis of their distribution between a gas and either a solid or a liquid. When a solid is used as the stationary phase (i.e. gas–solid chromatography), molecules adsorb onto the surface. This is therefore a form of adsorption chromatography. Ir clinical chemistry, gas–liquid chromatography is more widel used and it is this that will be described.

In this technique, the stationary phase is a non-volatile liquic (e.g. a polyethylene glycol or a silicone oil) bound onto particle of an inert solid (e.g. a diatomaceous earth). The sampl molecules are distributed between this liquid and the gas Gas–liquid chromatography is therefore a form of partitio chromatography. The stationary phase is packed into a colum which is held in a high temperature oven. Carrier gas, such a nitrogen or helium, is passed through the column. When th sample is injected it is immediately vaporized because of the hig temperature, and the molecules become distributed between th liquid and the gas. Because of the passage of the gas, th molecules are eventually resolved, depending on their distrib tion, between the liquid and the gas. The components whic emerge at the end of the column pass through a detector whic records their passage in the form of a peak on a chart recorde Several types of detector can be used and these incluc katharometers, electron capture detectors and flame ionizatio detectors.

The high boiling points of some compounds means that th must be converted into a lower boiling point form before th can be separated by this technique. Many compounds a therefore converted to more volatile derivatives, e.g. silyl or acetyl compounds, prior to being injected on the column.

Gas chromatography is a sensitive technique for the detection and quantitation of many substances in clinical chemistry. Lipids, organic acids, drugs, steroids and alcohol are some of the compounds which can be estimated by this method.

Further reading: (1971). Special issue. Gas chromatography in clinical chemistry. *Clin. Chem. Acta,* **34,** 129

Street, H.V. (1969). The use of gas–liquid chromatography in clinical chemistry. In Bodansky, O. and Stewart, C.P. (eds.) *Advances in Clinical Chemistry*. Vol. 12, p. 217 (New York: Academic Press)

GASTRIC JUICE

Fluid secreted by the stomach containing hydrochloric acid, digestive enzymes (for example pepsinogen) and intrinsic factor which is necessary for vitamin B_{12} absorption. Gastric juice secretion is stimulated by gastrin and also by the vagus nerve which responds to the stimulus of food.

GASTRIC STIMULATION TEST

A test of the ability of the stomach to secrete hydrochloric acid in response to a suitable stimulant. These include insulin, histamine, pentagastrin, caffeine and alcohol. After administration of the stimulant, the gastric juice is collected by means of a tube passed down into the stomach and the acid content is determined by titration against standard alkali.

GASTRIN

A polypeptide hormone produced by the G cells of the stomach in response to food and which stimulates, among other things, gastric acid secretion. It consists of 17 amino acids although smaller and larger variants have been described ('little gastrin' and 'big gastrin'). In all gastrin derivatives, the same C-terminal tetrapeptide is responsible for its physiological actions. A synthetic pentapeptide derivative, 'pentagastrin', has been prepared which consists of the physiologically active part of the hormone.

Elevated serum levels

High serum gastrin levels are found in the Zollinger–Ellison syndrome when it is thought to arise in most cases from a pancreatic tumour.

Measurement

Although gastrin can be measured by biological assay it is usually measured by radioimmunoassay.

Further reading: general list of clinical textbooks

GAUCHER'S DISEASE

An inborn error of metabolism, characterized by the accumulation in the tissues of cerebrosides containing glucose residues. It is due to a deficiency of glucocerebrosidase which normally cleaves glucose from glucosyl ceramides. The disease is characterized by the accumulation of lipid-laden cells in the bone marrow and spleen.

Further reading: Brady, R.O. (1978). Glucosyl ceramide lipidosis: Gaucher's disease. In Stanbury, J.B., Wyngaarden, J.B. and Fredrickson, D.S. (eds.) *The Metabolic Basis of Inherited Disease*. 4th Edn., p. 731. (New York: McGraw-Hill)

GAUSSIAN DISTRIBUTION (NORMAL DISTRIBUTION)

This is the distribution in values of a variable (e.g. a histogram) in the form of a symmetrical bell-shaped curve, i.e. the arithmetic mean, mode and median coincide.

GEL FILTRATION

A form of chromatography which separates compounds on the basis of their molecular weight. It consists of particles either in a column or on a thin layer plate, which by virtue of their cross linkages, are porous to lower molecular weight species but exclude higher molecular weight compounds. Lower molecular weight compounds therefore run more slowly than the larger compounds and this results in their separation. Gel filtration can be used in the measurement of haptoglobins and in the diagnosis of macroglobulinaemia.

Further reading: general list of analytical textbooks

GEL IMMUNOFILTRATION

An immunochemical technique similar to immunoelectrophoresis, the difference being that instead of separating the proteins in the patients sample by electrophoresis, they are separated on a plate of dextran gel (e.g. Sephadex) which resolves the molecules on the basis of their molecular size. Addition of antiserum parallel and adjacent to the direction of filtration results in the formation of a series of precipitin lines. The technique can be used to investigate for example monomer and pentamer IgM in macroglobulinaemia.

Further reading: Grant, G.H. and Butt, W.R. (1970). Immunochemical methods in clinical chemistry. In Bodansky, O. and Stewart, C.P. (eds.) *Advances in Clinical Chemistry*. Vol. 13, p. 383. (New York: Academic Press)

GENERALIZED GANGLIOSIDOSIS

An inborn error of lipid storage characterized by the accumulation of ganglioside GM_1. It is due to a deficiency of the enzyme β-galactosidase which catalyses the removal of galactose from the GM_1 ganglioside molecule. The symptoms of the condition are somewhat similar to Hurler's syndrome but without the mucopolysaccharide excretion in the urine.

Further reading: O'Brien, J.S. (1978). The gangliosidoses. In Stanbury J.B., Wyngaarden, J.B. and Fredrickson, D.S. (eds.) *The Metabolic Basis of Inherited Disease*. 4th Edn., p. 841. (New York: McGraw-Hill)

GERHARDT'S TEST

A test which is used for the detection of ketone bodies in urine. It is based on the reaction of ferric chloride with acetoacetic acid to give a purple colour.

See also: **ketone bodies**

GIGANTISM

A condition that can result from excess growth hormone secretion in childhood usually from a pituitary adenoma.

See also: **human growth hormone**

GILBERT'S DISEASE

A relatively harmless form of congenital hyperbilirubinaemia possibly caused by failure of bilirubin to be transported into the liver cell. Elevated levels of unconjugated bilirubin are found in the plasma.

Further reading: Schmid, R. and McDonagh, A.F. (1978). Hyperbilirubinaemia. In Stanbury, J.B., Wyngaarden, J.B. and Fredrickson, D.S. (eds.) *The Metabolic Basis of Inherited Disease*. 4th Edn., p. 1221. (New York: McGraw-Hill)

GLOBIN

The protein which is combined with haem to form haemoglobin.

See also: **haemoglobin**

GLOBOID LEUKODYSTROPHY

See: **Krabbe's disease**

GLOBULINS

These are proteins which are soluble in dilute salt solutions but are only sparingly soluble or insoluble in water (unlike albumin which is soluble in both water and salt solutions). The globulins have been subdivided into euglobulins which are insoluble in water and the pseudoglobulins which are sparingly soluble in water.

However the globulins of human serum are usually classified into four major categories on the basis of their electrophoretic behaviour.

(1) α_1-Globulins. The major constituents of this fraction are α antitrypsin, and α_1-acid glycoprotein.

(2) α_2-Globulins. Within this group are the haptoglobins caeruloplasmin and α_2-macroglobulin.

(3) β-Globulins. Transferrin and the C3 component of complement are in this group.

(4) γ-Globulins. This fraction contains the immunoglobulins IgG, IgA, IgM, IgD and IgE.

Measurement

Immunochemical techniques are available for the determination of many of the individual globulins. The globulin fraction as a whole can be determined by procedures based on the Hopkins–Cole reaction for tryptophan since globulins, unlike albumin contain tryptophan residues in their structure.

See also: **proteins**

GLOMERULAR FILTRATION RATE (GFR)

The rate at which plasma is filtered at the renal glomeruli. It can be determined by measuring the clearance of a substance which is filtered freely at the glomerulus and is neither secreted nor reabsorbed by the tubules. Such substances include inulin, mannitol and ^{51}Cr-labelled EDTA. Creatinine is secreted in small amounts by the renal tubules and this gives slightly higher clearance values than the other substances although this still approximates to the glomerular filtration rate.

See also: **creatinine, clearance**

GLUCAGON

A polypeptide hormone secreted by the α cells of the islets of Langerhans in the pancreas. Glucagon is secreted in response to hypoglycaemia and it raises the blood sugar level by stimulating liver glycogenolysis and gluconeogenesis. A curious effect of glucagon is nevertheless the stimulation of insulin secretion.

Increases in serum levels

The only situation in which glucagon estimations are useful is in the diagnosis of rare glucagon secreting tumours of the pancreas.

Measurement

The hormone is usually measured by radioimmunoassay.

Further reading: Bloom, S.R. (1975). Glucagon. *Br. J. Hosp. Med.,* **13**, 150

GLUCAGON TOLERANCE TEST

A test which is particularly useful in the diagnosis of insulinoma and some types of glycogen storage diseases, especially types I and

III. It consists of the intravenous injection of glucagon followed by the collection of blood samples at intervals for glucose estimation. A normal response is a moderate elevation of the blood glucose level. In patients with an insulinoma, there is an initial hyperglycaemia followed by severe hypoglycaemia due to the excessive release of insulin. In patients with types I and III glycogen storage diseases the blood glucose level alters very little.

See also: **glycogen storage diseases, insulinoma**

GLUCOCORTICOIDS

Steroid hormones produced by the adrenal cortex which influence carbohydrate metabolism. The major glucocorticoids are cortisol and corticosterone. Their actions are the opposite of those of insulin in that they increase gluconeogenesis and decrease the peripheral utilization of glucose, thus tending to raise the blood sugar level. They also increase protein breakdown and stimulate lipolysis.

See also: **cortisol**

GLUCOSE

A key molecule in carbohydrate metabolism. It is formed both a a result of the digestion of complex carbohydrates and as a resul of *de novo* synthesis within the body (gluconeogenesis). It can b synthesized into glycogen, the storage form of glucose (glycogenesis) or it can be catabolized via the glycolytic an tricarboxylic acid pathways to provide energy for the body' needs. It is also metabolized by the hexose monophosphate shur (pentose shunt) to provide pentose sugars for nucleic acid syr thesis and NADPH which is involved in many biosynthet reactions.

Hormonal control of blood glucose levels

Several hormones are involved in blood glucose homeostasi:

(1) Insulin. This lowers blood glucose levels by:
   (a) Promoting entry of glucose into cells, and
   (b) Stimulating glycogenesis and inhibiting gluc neogenesis.

(2) Glucagon. This raises blood sugar levels by:
  (a) Stimulating liver glycogenolysis, and
  (b) Stimulating gluconeogenesis.

(3) Growth hormone. This tends to raise the blood sugar level by opposing the actions of insulin on carbohydrate metabolism.

(4) Glucocorticoids. These oppose the actions of insulin and thus tend to raise the blood sugar level.

(5) Adrenaline. This also has anti-insulin effects, increasing glycogenolysis and thereby raising the blood sugar level.

Elevated blood glucose levels

Hyperglycaemia is found in diabetes mellitus when insufficient insulin is present. This may be a primary condition or secondary to other conditions, e.g. when insulin antagonists are present in excess as in Cushing's disease (due to excess glucocorticoids), acromegaly (due to excess growth hormone) or phaeochromocytoma (due to excess adrenaline).

Decreased blood glucose levels

The causes of hypoglycaemia can be divided into three categories:

(1) Drug-induced hypoglycaemia, e.g. overdosage of exogenous insulin or oral hypoglycaemic drugs.

(2) Reactive hypoglycaemia. This is when hypoglycaemia results from the ingestion of a particular substance, e.g.

  (a) Functional hypoglycaemia as a result of sensitivity to glucose.
  (b) It occurs after meals in patients with a gastrectomy.
  (c) In leucine sensitivity.
  (d) In alcohol induced hypoglycaemia.
  (e) In hereditary fructose intolerance.
  (f) In galactosaemic patients after milk (which contains lactose) ingestion.

(3) Fasting hypoglycaemia. Among the conditions in which this occurs are:

  (a) Excess endogenous insulin secretion as in insulinoma.
  (b) When there is severe liver damage.
  (c) Glycogen storage disease.

(d) Pituitary or adrenal insufficiency, e.g. Addison's disease.
(e) In many malignant tumours.
(f) Idiopathic hypoglycaemia which can occur in infants.

Glycosuria

The finding of glucose in urine indicates that hyperglycaemia is present except in those subjects with a low renal threshold for glucose.

Measurement of blood glucose

(1) Glucose can be estimated by its ability to reduce the yellow ferricyanide ion to the colourless ferrocyanide ion in hot alkaline solutions. This technique therefore involves measuring a decrease in colour. A disadvantage of the method is that other reducing substances such as creatinine or uric acid interfere with the assay giving falsely high glucose values. This method can be used on continuous flow analysers.

(2) Another category of methods makes use of the ability of glucose to reduce cupric ion to cuprous ion resulting in the formation of cuprous oxide. Among the methods are:

(a) The Folin–Wu procedure. A protein-free filtrate is prepared using tungstic acid. Glucose in the filtrate reduces cupric ion to cuprous ion and the latter in turn reduces phosphomolybdic acid to molybdenum blue which can then be estimated colorimetrically. Thi method is now little used.

(b) The neocuproine method. In this technique th cuprous ion forms a coloured complex wit neocuproine and this can be measured colorimetri cally. This method can be used on continuous flo instruments. A disadvantage is however interferenc by other compounds, e.g. uric acid.

(c) The Somogyi–Nelson procedure is also based o copper reduction. Proteins are removed by bariu hydroxide and zinc sulphate. Like the Folin–W procedure, this method is now virtually obsolete.

(3) Glucose oxidase methods. This enzyme catalyzes the rea tion

$$\text{glucose} + O_2 + H_2O \longrightarrow \text{gluconic acid} + H_2O_2$$

The enzyme is highly specific for glucose and therefore gives results much nearer to the true glucose value than the previous two categories of methods. The glucose oxidase reaction can be followed in one of two ways:

(a) By measuring the rate of oxygen consumption, for instance in the Beckman glucose analyser which uses a polarographic oxygen electrode.

(b) By using the hydrogen peroxide formed in the above reaction to oxidize a chromogenic oxygen acceptor, a reaction catalysed by the enzyme peroxidase. Among the oxygen accepting dyes which can be used are *o*-tolidine, *o*-dianisidine (although these are carcinogenic) and 4-aminophenazone-phenol as in the Trinder method. Glucose oxidase methods can be adapted for use on continuous flow machines.

A technique which is based on the glucose oxidase reaction is the Dextrostix (Ames), in which a drop of blood is placed on a plastic strip. The intensity of the eventual colour formed is proportional to the glucose concentration.

(4) Hexokinase methods. This enzyme catalyses the reaction

$$\text{glucose} + \text{ATP} \longrightarrow \text{glucose-6-phosphate} + \text{ADP}$$

Glucose-6-phosphate dehydrogenase is included in the reaction mixture. This converts the glucose-6-phosphate formed to 6-phosphogluconate, a reaction in which NADP is reduced to NADPH, and this can be followed spectrophotometrically.

(5) Condensation reactions. These are based on the reaction of glucose in hot acetic acid with various aromatic amines, e.g. aniline or *o*-toluidine to produce coloured derivatives. Other sugars can also react, giving falsely high values.

Detection of glucose in urine

(1) Many methods are based on the ability of glucose to reduce Benedict's reagent which contains cupric ion complexed with citrate in alkaline solution. However other reducing substances in urine may also give false positives.

(2) Glucose can be detected with Fehling's solution, another reaction based on copper reduction. Other reducing substances also interfere with this test.

(3) Clinistix strips (Ames) are usually more specific for glucose since they are based on the glucose oxidase reaction.

See also: **diabetes mellitus**

Further reading: Jarrett, R.J. (1971). Blood glucose homeostasis and its disorders. *Bri. J. Hosp. Med.*, **6**, 499.
General list of analytical and clinical textbooks.

GLUCOSE OXIDASE

An enzyme which can be used for the detection and quantitation of glucose in biological fluids. It catalyses the reaction

$$\text{glucose} + O_2 + H_2O \longrightarrow \text{gluconic acid} + H_2O_2$$

The reaction can be quantified by measuring the amount of oxygen consumed by the reaction (e.g. a Beckman glucose analyser) or by determination of the hydrogen peroxide formed by means of peroxidase and an oxygen accepting dye.

See also: **glucose**

GLUCOSE-6-PHOSPHATASE

An enzyme which catalyses the interconversion of glucose and glucose-6-phosphate. A deficiency of this enzyme occurs in glycogen storage disease type I (Von Gierke's disease).

See also: **glycogen storage diseases**

GLUCOSE-6-PHOSPHATE DEHYDROGENASE (G-6-PD) DEFICIENCY

G-6-PD is one of the enzymes of the pentose phosphate shunt, a pathway resulting in the generation of NADPH which has a role in maintaining the integrity of the red cell membrane. Various forms of red cell G-6-PD deficiency can occur, and these can result in a haemolytic anaemia. In some cases, haemolytic episodes are precipitated by drugs particularly some antimalarial drugs such as primaquine. Another variant of the condition is favism, where a haemolytic episode is triggered by ingestion of fava beans from the bean plant *Vicia fava*.

Measurement

The enzyme is usually measured in red blood cell haemolysates, using glucose-6-phosphate as the substrate. This is oxidized to 6-

phosphogluconolactone with a simultaneous reduction of NADP to NADPH which can be followed spectrophotometrically.

Further reading: Beutler, E. (1978). Glucose-6-phosphate dehydrogenase deficiency. In Stanbury, J.B. Wyngaarden, J.B. and Fredrickson, D.S. (eds) *The Metabolic Basis of Inherited Disease*. 4th Edn., p. 1430. (New York: McGraw-Hill)

GLUCOSE TOLERANCE TESTS

These are *in vivo* tests used for the investigation of carbohydrate metabolism. They consist of giving an oral glucose load (usually 50 g for adults) to fasting subjects followed by the collection of blood and urine samples at intervals for up to two hours or sometimes longer. There are several types of abnormal curve.

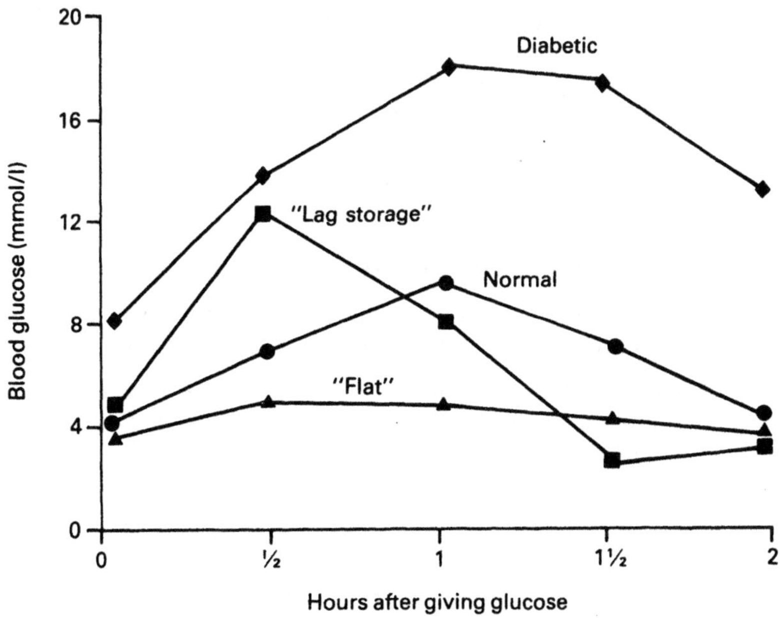

(a) Diabetic type, i.e. high blood sugar levels often with glycosuria.

(b) A 'Flat' curve. This can occur in malabsorption, in Addison's disease (due to low glucocorticoid levels) or in growth hormone deficiency.

(c) A 'Lag Storage' curve in which low blood sugar levels are found in the later specimens. The hypoglycaemia may be due to an overproduction of insulin. It can occur in otherwise normal subjects or in patients with a gastrectomy.

(d) Renal glycosuria. In this case the blood sugar levels are normal but glycosuria is found due to a low renal threshold.

Variations in the glucose tolerance test

(1) Intravenous glucose tolerance test. This is used for investigating glucose metabolism when abnormal carbohydrate absorption is present.

(2) Cortisone stressed glucose tolerance test. This is a test for latent diabetes which can be performed if a conventional glucose tolerance test has been found to be normal. It consists of giving cortisone, followed by an oral glucose load. An abnormal glucose tolerance curve would suggest latent diabetes.

See also: **diabetes mellitus, glucose**

GLUTAMATE DEHYDROGENASE

An enzyme found in liver cell mitochondria, the measurement of which in the serum can be indicative of hepatocellular disease.

Further reading: Wilkinson, J.H. (1976) *The Principles and Practice of Diagnostic Enzymology*. (London: Edward Arnold)

GLUTAMATE-OXALOACETATE TRANSAMINASE (GOT)

See: **aspartate aminotransferase**

GLUTAMATE-PYRUVATE TRANSAMINASE (GPT)

See : **alanine aminotransferase**

GLUTAMINE

A compound formed by the combination of ammonium with glutamate. Most glutamine is metabolized by the liver to form glutamate and the ammonium nitrogen is converted to urea. In severe liver disease, ammonia is not removed and this leads to increased glutamine synthesis by the brain. Thus increased blood glutamine levels are often found in liver disease.

Glutamine is hydrolysed in renal tubular cells by the enzyme glutaminase and this is considered to have a role in the renal buffering mechanisms.

γ-GLUTAMYL TRANSFERASE (γ-GLUTAMYL TRANSPEPTIDASE, γGT)

An enzyme which catalyses the transfer of of a γ-glutamyl group from a peptide to an acceptor molecule. The enzyme present in blood originates mainly from the hepatobiliary system.

Increases in serum levels

High serum levels of the enzyme are found in all forms of liver disease but especially high levels are found in biliary obstruction and it is therefore an important indicator of cholestasis. The enzyme has also been found to be a sensitive indicator of alcohol abuse as high levels occur in the serum of heavy drinkers.

Measurement

γ-Glutamyl transferase can be measured using γ-glutamyl-*p*-nitroanilide as the substrate and glycylglycine as the accepting molecule:

$$\begin{array}{ccc} \gamma\text{-glutamyl-}p\text{-nitroanilide} & \longrightarrow & \gamma\text{-glutamylglycylglycine} \\ +\ \text{glycylglycine} & & +\ p\text{-nitroaniline} \end{array}$$

The increase in the absorbance at 405 nm due to the formation of *p*-nitroaniline is measured spectrophotometrically. Another substrate which can be used is γ-glutamyl-3-carboxy-4-nitroanilide.

Further reading: Rosalki, S.B. (1975). Gamma-glutamyl transpeptidase. In Bodansky, O. and Latner, A.L. (eds.) *Advances in Clinical Chemistry*. Vol. 17, p. 53. (New York: Academic Press)

Wilkinson, J.H. (1976). *The Principles and Practice of Diagnostic Enzymology.* (London: Edward Arnold)

GLUTATHIONE

A tripeptide containing a sulphydryl group which has a role in maintaining the integrity of erythrocytes by virtue of it being an important regulator of oxidation–reduction reactions. It is known to interfere in some methods of blood sugar estimation.

GLUTATHIONE REDUCTASE

The enzyme which replenishes the red blood cell stores of reduced glutathione. It catalyses the following reaction using NADP and flavin adenine dinucleotide (FAD) as cofactors:

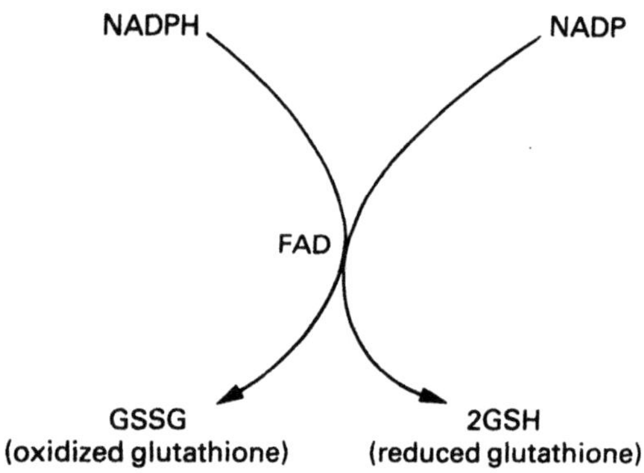

It can be measured spectrophotometrically by following the oxidation of NADPH at 340 nm.

The enzyme can be measured in red blood cells to assess the riboflavin status of the body. Riboflavin is a component of the FAD coenzyme. When the body's riboflavin status is good, the enzyme is saturated with FAD and addition of exogenous FAD to the assay mixture will result in little or no increase in activity. On the other hand, if there is inadquate riboflavin intake, addition of FAD causes a marked activation of the enzyme.

See also: **riboflavin**

Further reading: Wilkinson, J.H. (1976) *The Principles and Practice of Diagnostic Enzymology.* (London: Edward Arnold)

GLUTATHIONE STABILITY TEST

A test used for the determination of red cell glucose-6-phosphate dehydrogenase activity. It consists of incubating red cells with acetylphenylhydrazine followed by measurement of the glutathione levels. Little change occurs with normal erythrocytes but with glucose-6-phosphate dehydrogenase deficient cells a marked reduction in the glutathione level occurs.

See also: **glucose-6-phosphate dehydrogenase deficiency**

GLUTEN SENSITIVE ENTEROPATHY

See: **coeliac disease**

GLUTETHIMIDE (DORIDEN)

A nonbarbiturate hypnotic. Glutethimide poisoning causes symptoms similar to those of barbiturate poisoning except that they may be more severe. It can be detected in body fluids by ultra violet absorption spectrophotometry after an initial extraction.

Further reading: Meade, B.W. *et al.* (1972). Technical Bulletin No. 24. Simple tests to detect poisons. *Ann. Clin. Biochem.*, **9**, 35

GLYCINE

An amino acid found in elevated amounts in blood and urine in the inborn errors of metabolism, non ketotic hyperglycinaemia, (which is possibly due to a deficiency in the enzyme glycine decarboxylase) and propionic acidaemia (ketotic hyperglycinaemia). It is also found in increased amounts in urine along with proline and hydroxyproline in another inborn error, familial iminoglycinuria. This is due to defective membrane transport of these amino acids.

[^{14}C]GLYCOCHOLIC ACID BREATH TEST

An *in vivo* test for the detection of bacterial overgrowth in the small bowel or terminal ileum. It consists of the oral administration to the patient of [^{14}C]glycocholic acid, where the radioactive label is part of the glycine portion of the molecule. The bacteria are able to split [^{14}C]glycocholic acid into [^{14}C]glycine and cholic acid. The [^{14}C]glycine is absorbed and then metabolized by the tissues to yield $^{14}CO_2$ which is eliminated by the lungs. The bacteria themselves may also degrade the [^{14}C]glycine to give $^{14}CO_2$ which is absorbed and eliminated by the lungs. The $^{14}CO_2$ in the expired air is measured by using a suitable trapping agent and counting the radioactivity. The presence of increased amounts of $^{14}CO_2$ in expired air is suggestive of bacterial overgrowth in the upper part of the alimentary tract.

Further reading: Editorial (1975). Breath tests in gastroenterology. *Lancet*, **2**, 163

GLYCOGEN

A branched chain polysaccharide composed of glucose units. It functions as a storage form of carbohydrate in muscle, liver and other tissues. The glucose units are joined together in α-1,4 linkages except at the branch points where there is an α-1,6 linkage.

A simplified version of the synthesis and breakdown is:

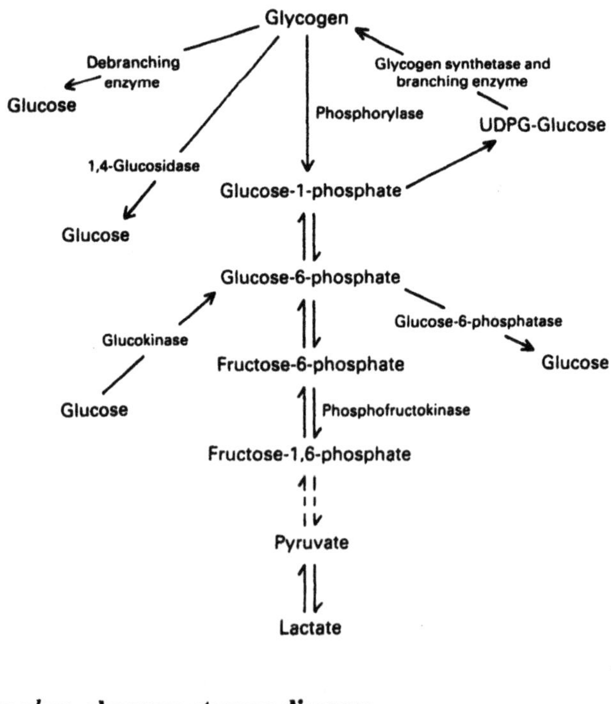

See also: **glycogen storage diseases**

GLYCOGEN STORAGE DISEASES

A group of inborn errors of metabolism in which one of the enzymes involved in the degradation or synthesis of glycogen is deficient. This results in the formation of an abnormal type of glycogen or the accumulation of normal glycogen. There are at least eight types of glycogen storage disease and these are named Type I to Type VIII.

Type I (Von Gierke's disease)

This is the commonest form of glycogen storage disease and it is due to a deficiency of the enzyme glucose-6-phosphatase. This converts glucose-6-phosphate to glucose, one of the steps in glycogen breakdown. Glycogen accumulates in the liver and hepatomegaly results. Hypoglycaemia is found as a result of the inability to utilize glycogen. Two biochemical tests which are useful in diagnosing the condition are:

(1) A glucagon tolerance test. This normally stimulates glycogenolysis resulting in a rise in blood sugar but this does not occur in Type 1 glycogen storage disease.

(2) A galactose or fructose tolerance test normally results in a rise in blood sugar levels as a result of their conversion in the liver by a series of enzymes including glucose-6-phosphatase. In Type I glycogen storage disease this does not occur.

Type II (Pompe's disease)

This is due to a deficiency of one of the enzymes of glycogen degradation, α-1,4-glucosidase. Accumulations of glycogen are found in many tissues.

Type III (Cori's disease)

This is due to the deficiency of the enzyme amylo-1,6-glucosidase, the enzyme which 'debranches' glycogen. As a result, a glycogen having abnormally short branches accumulates in the tissues.

Type IV

In this condition there is a deficiency of amylo-1,4→1,6-transglucosylase, the enzyme involved in the formation of the branch points of glycogen. An abnormal glycogen with long outer branches accumulates.

Type V (McArdle's disease)

This is due to the absence of muscle phosphorylase activity. Phosphorylase is one of the glycolytic enzymes.

Type VI (Hers' disease)

This is due to the absence of liver phosphorylase.

Type VII

The deficient enzyme in this case is phosphofructokinase.

Type VIII

The deficient enzyme in this case is phosphorylase b-kinase, the enzyme involved in the activation of phosphorylase.

Further reading: Howell, R.R. (1978). The glycogen storage diseases. In Stanbury, J.B. Wyngaarden, J.B. and Fredrickson, D.S. (eds.) *The Metabolic Basis of Inherited Disease*. 4th Edn., p. 137. (New York: McGraw-Hill)

GLYCOPROTEINS

Proteins linked covalently to carbohydrates. (Those containing a small percentage of carbohydrate tend to be known as glycoproteins whereas those with larger percentages are referred to as mucoproteins.) Examples of glycoproteins are fibrinogen and transferrin. Serum levels of glycoproteins are raised in many inflammatory conditions.

GLYCOSYLATED HAEMOGLOBINS (FAST HAEMOGLOBINS)

These are formed by the reaction of glucose with normal adult haemoglobin. Their concentration in blood is proportional to the average blood glucose level. Their measurement can therefore give an indication of diabetic control over a period of several weeks.

They can be measured spectrophotometrically after separation from normal haemoglobin by ion exchange resins.

Further reading: Editorial (1977). Glycosylated haemoglobins and disease. *Lancet*, **2**, 22

Editorial (1978). Glycosylated haemoglobin and diabetic control. *Br. Med. J.*, **1**, 1373

GMELIN'S TEST

An old test for the detection of urinary bilirubin. It consists of layering the urine onto concentrated nitric acid. A blue colour at the interface of the two liquids indicates the presence of bilirubin.

GODFRIED'S TEST

A test for bile pigments in urine based on the Van den Bergh diazo reaction.

GOLD

Gold salts are used occasionally in the treatment of rheumatoid arthritis. Serum levels of gold can be measured by atomic absorption spectrophotometry.

Further reading: Editorial. (1979) Fifty years of gold in rheumatoid arthritis. Br. Med. J., 1., 289

GONADOTROPHINS

These are hormones which influence the function and maturation of the ovary or testis. The main ones are follicle-stimulating hormone (FSH) and luteinizing hormone (LH), both of which are secreted by the anterior pituitary gland.

Human chorionic gonadotrophin (HCG) is structurally similar to LH and has many of its actions. It is secreted by the placenta and by trophoblastic tumours.

See also: **follicle-stimulating hormone, human chorionic gonadotrophin, luteinizing hormone**

GOODPASTURE'S SYNDROME

An autoimmune disease in which autoantibodies acting against the basement membrane of glomerulus and lung are found. It is characterized by recurrent haemorrhages into the lung associated with proliferative glomerulonephritis.

Further reading: Roitt, I.M. (1977). *Essential Immunology.* 3rd Edn., (Oxford, London, Edinburgh, Melbourne: Blackwell Scientific Publications)

GOUT

This condition is caused by the precipitation of uric acid cystals in joints and other tissues as a result of hyperuricaemia. It can be a primary condition or secondary to another disease.

Primary gout

This is due to increased uric acid production from purines, possibly due to overactivity of one of the enzymes in the pathway. Reduced renal secretion of urate may also be a factor. There are three forms of treatment:

(1) Reducing dietary purine intake.

(2) Increasing the renal excretion of urate using uricosuric drugs such as salicylate or probenecid.

(3) Reducing uric acid production with drugs such as allopurinol which inhibit xanthine oxidase, an enzyme involved in the formation of uric acid.

Secondary gout

High plasma uric acid levels can result from:

(1) Increased turnover of nucleic acids, e.g. rapidly growing malignancies or tissue damage.

(2) Reduced excretion of uric acid, e.g. renal disease.

Diagnosis of gout

(1) Fluid from affected joints can be examined microscopically for uric acid crystals.

(2) By finding elevated blood uric acid levels.

See also: **uric acid**

Further reading: Wyngaarden, J.B. and Kelley, W.N. (1978). Gout. In Stanbury, J.B., Wyngaarden, J.B. and Fredrickson, D.S. (eds.) *The Metabolic Basis of Inherited Disease*. 4th Edn. p. 916 (New York: McGraw-Hill)

GRAVE'S DISEASE

The commonest form of hyperthyroidism. It is probably autoimmune in origin, being due to the presence of long-acting thyroid stimulator (LATS), an immunoglobulin capable of stimulating the thyroid gland.

See also: **hyperthyroidism**

GROWTH HORMONE

See: **human growth hormone**

GROWTH HORMONE RELEASING FACTOR (GHRF)

A hypothalamic polypeptide hormone which controls the release of growth hormone from the anterior pituitary.

See also: **human growth hormone**

GROWTH HORMONE SUPPRESSION TEST

A test used in the diagnosis of acromegaly. It consists of giving glucose to the patient followed by the collection of blood samples for growth hormone estimation at half-hour intervals for up to two hours. In normal subjects, glucose causes a suppression of growth hormone levels but in acromegalic patients no suppression occurs.

See also: **acromegaly**

GUANASE (GUANINE DEAMINASE)

An enzyme present in liver and other organs which converts guanine to xanthine. It can be measured in serum as an indication of liver disease.

Further reading: Wilkinson, J.H. (1976). *The Principles and Practice of Diagnostic Enzymology*. (London: Edward Arnold)

GUNZBURG'S TEST

A test for the detection of hydrochloric acid in gastric aspirates based on the formation of a red colour with phloroglucinol and vanillin.

GUTHRIE TEST

This is a microbiological procedure which can be used for the detection of elevated blood phenylalanine levels. It is based on the ability of phenylalanine to overcome the effects of β-2-thienylalanine, a metabolic antagonist to the growth of a strain of *Bacillus subtilis*. The test consists of placing filter paper discs impregnated with neonatal blood on an agar plate containing growth medium and metabolic antagonist. The presence of bacterial growth around the disc indicates elevated blood phenylalanine levels, suggesting the infant has phenylketonuria.

See also: **phenylketonuria**

H

HABA (2-(4′-HYDROXYAZOBENZENE)-BENZOIC ACID)

A dye which binds to albumin and which can therefore be used for its estimation.

HAEM

A tetrapyrolle containing iron in the ferrous state. It is linked to the protein globin to form haemoglobin.

HAEMAGGLUTINATION TECHNIQUES

These are tests based on the aggregation of red blood cells and which are used for the detection of antigens on their surfaces. Agglutination occurs because of the cross linking of cells by antibodies directed against surface antigens. There are several different types:

(1) Direct agglutination reactions which are used in blood banking to test for erythrocyte blood group surface antigens. Specific blood group antisera are used.

(2) Indirect or passive haemagglutination in which a foreign antigen, or occasionally an antibody, is coated into the surface of the erythrocytes. In the presence of antibody (or antigen) agglutination occurs.

(3) Haemagglutination inhibition. This is based on the ability of certain antigens to inhibit haemagglutination of coated cells by antibody.

Further reading: General list of analytical textbooks

HAEMATIN

Haem which has been oxidized and contains iron in the ferric state. In plasma, haematin can combine with albumin to form methaemalbumin. Haematin is released from red cells in patients with intravascular haemolysis and in methaemoglobinaemia.

HAEMOCHROMATOSIS

A genetically determined condition of iron overloading in which iron accumulates in many of the organs of the body including liver, heart, bone and spleen. It is due to the increased intestinal absorption of iron. It can present clinically as liver cirrhosis with diabetes mellitus and skin pigmentation and hence it is sometimes known as bronzed diabetes. Patients usually have a high plasma iron level with a low total iron binding capacity. Iron can usually be found histologically in liver biopsy specimens. Chelating agents, such as desferrioxamine which bind iron and are subsequently excreted in the urine, are used in the treatment of the condition.

Further reading: Pollycove, M. (1978). Haemochromatosis. In Stanbury, J.B., Wyngaarden, J.B. and Fredrickson, D.S. (Eds.) *The Metabolic Basis of Inherited Disease*. 4th Ed., p. 1127. (New York: McGraw-Hill)

HAEMOGLOBIN

A red cell protein comprised of four protein subunits (globin) each of which consists of a polypeptide chain linked to haem. In normal haemoglobin there are two α chains and two β chains. It is involved in the transport of oxygen in blood and also functions in acid–base homeostasis.

Haemoglobin levels in whole blood

When the haemoglobin concentration is below normal, the patient is said to have anaemia. Measurement of haemoglobin concentration in whole blood usually falls within the province of the haematology laboratory and for this reason, discussion of this aspect will be limited.

Increased plasma levels of haemoglobin

Increases in plasma haemoglobin are seen when there is intravascular haemolysis, e.g. haemolytic anaemias. The haemoglobin can bind to haptoglobin to prevent its loss in the urine. If the binding capacity of the haptoglobin is exceeded however, haemoglobinuria results.

Haemoglobinuria

This can occur as a result of intravascular haemolysis (see above) or as a result of bleeding within the urinary tract, when intact erythrocytes can appear in the urine.

Haemoglobin in faeces

This indicates bleeding within the gastrointestinal tract. For a more detailed discussion see **occult blood.**

Haemoglobin measurement

Although the measurement of haemoglobin in whole blood is usually carried out by the haematology laboratory, the clinical chemistry laboratory may be asked to detect it in plasma, urine and faeces. In plasma and urine it can be identified by its characteristic absorption spectrum. It can also be detected in all three types of speciment by its peroxidase ability. Haemoglobin catalyses the reduction of hydrogen peroxide to water and at the same time a dye (e.g. guaiac) is oxidized to a coloured form.

Further reading: Textbooks of haematology.
General list of clinical and analytical textbooks.

HAEMOGLOBINOPATHIES

A group of inherited disorders in which there is an amino acid substitution or deletion in the haemoglobin molecule. Such alterations can result in changes in the solubility of haemoglobin and its oxygen-carrying capacity in the blood. Many of the haemoglobinopathies are harmless, but some are not, the most common being sickle cell anaemia.

See also: **sickle cell anaemia**

Further reading: Winslow, R.M. and Anderson, W.F. (1978). The haemoglobinopathies. In Stanbury, J.B., Wyngaarden, J.B. and Fredrickson, D.S. (eds.) *The Metabolic Basis of Inherited Disease.* 4th Ed., p. 1465. (New York: McGraw-Hill)
Huisman, T.H.J. (1972). Normal and abnormal human haemoglobins. In Bodansky, O. and Latner, A.L. (eds.) *Advances in Clinical Chemistry.* Vol. 15, p. 150. (New York: Academic Press)

HAEMOLYTIC ANAEMIAS

Anaemia due to intravascular lysis of erythrocytes. Among the causes of haemolytic anaemia are:

(1) There may be a hereditary abnormality of the red cell, e.g. sickle cell anaemia and glucose-6-phosphate dehydrogenase deficiency.

(2) As a result of infections, e.g. haemolytic streptococcus.

(3) In mismatched transfusions.

(4) As a result of poisoning, e.g. by arsine.

(5) In rhesus haemolytic disease when there is destruction of fetal red cells by maternal antibodies.

(6) Autoimmune haemolytic anaemias.

Biochemical features

Among the biochemical features of haemolytic anaemias are:

(1) Elevated bilirubin levels in the serum, mainly of the unconjugated variety.

(2) Low serum haptoglobin levels.

(3) Methaemalbumin may be present.

Further reading: Emerson, P.M. (1971). Haemolytic anaemias: aetiology and diagnosis. *Br. J. Hosp. Med.,* **6**, 607

Flaherty, T. and Geary, C.G. (1979). Autoimmune haemolytic anaemia. *Br. J. Hosp. Med.,* **22**, 334

HAEMOPEXIN

A plasma protein which can bind haem.

HAEMOSIDERIN

A storage form of iron consisting of small iron oxide granules. It can accumulate in the body when the iron present exceeds the capacity of the body to store it as ferritin. In cases where there is iron overload, haemosiderin is deposited in the tissues and excreted in the urine. It can be detected in urine and tissues by direct microscopy after staining with potassium ferrocyanide.

See also: **iron**

HAGEDORN–JENSEN METHOD

A method of blood sugar estimation based on the ability of glucose to reduce ferricyanide to ferrocyanide. The ferricyanide remaining is determined iodometrically.

See also: **glucose**

HAMBURGER SHIFT

See also: **chloride shift**

HANTZSCH REACTION

A condensation reaction in which formaldehyde reacts with acetylacetone and ammonium ion to produce diacetyl lutidine which is coloured yellow. It is used in the estimation of triglycerides after the glycerol of the triglyceride molecule has been oxidized to formaldehyde. It can also be used for the estimation of uric acid in conjunction with the enzyme uricase.

See also: **triglycerides, uric acid**

HAPTEN

A small molecule, such as a drug or steroid hormone, that by itself cannot stimulate the production of antibodies, but which can do so when combined with a large protein molecule. The antibodies formed are then capable of interacting with the small molecule.

HAPTOGLOBINS

These are α_2-globulins which can combine with plasma free haemoglobin (resulting from red cell breakdown). Unlike free haemoglobin, these complexes are too large to pass through the glomerulus and by this means iron is conserved by the body. The haemoglobin–haptoglobin complex is catabolized at a faster rate than haptoglobin alone. Thus low serum levels of haptoglobins are found when there is intravascular haemolysis. A congenital deficiency of haptoglobin has also been described. Raised serum levels are found in a variety of inflammatory states.

Measurement

(1) In some methods, an excess of haemoglobin is added to the serum in order to saturate the haptoglobin. The complex can be separated from free haemoglobin by electrophoresis or gel filtration and the amount of haemoglobin binding to haptoglobin can be determined.

(2) Some methods are based on the fact that the haemoglobin–haptoglobin complex has a greater peroxidase activity than haemoglobin alone.

(3) Immunochemical methods such as radial immunodiffusion can be used, but a major difficulty is the different genetic forms of haptoglobin which have differing affinities for the antibody.

Further reading: General list of analytical and clinical textbooks

HARTNUP DISEASE

An inborn error in which there is defective intestinal and renal transport of neutral amino acids, one of these being tryptophan. This amino acid is normally converted to the vitamin, nicotinamide. The clinical features of Hartnup disease are similar to the nicotinamide deficiency disease, pellagra, being due to the low amounts of tryptophan available for nicotinamide synthesis. The disease can be diagnosed by the presence of large amounts of indole compounds in the urine, which result from the action of gut bacteria on the unabsorbed dietary tryptophan.

Further reading: Jepson, J.B. (1978). Hartnup disease. In Stanbury, J.B., Wyngaarden, J.B. and Fredrickson, D.S. (eds.) *The Metabolic Basis of Inherited Disease*. 4th Ed., p. 1563. (New York: McGraw-Hill)

HARTRIDGE REVERSION SPECTROSCOPE

An instrument for the direct viewing of absorption spectra of solutions. The optics of the instrument are designed such that two absorption spectra are produced, one of which appears above the other but in the reversed position. By aligning the two spectra, the absorbing wavelength can be read off a vernier scale. The instrument can be used for the identification of haemoglobin derivatives such as carboxyhaemoglobin.

HASHIMOTO'S DISEASE

A form of hypothyroidism which is due to the destruction o thyroid tissue by thyroid autoantibodies.

See also: **hypothyroidism**

HAY'S TEST

A test for the presence of excess bile salts in urine which consists c sprinkling flowers of sulphur on the surface. If bile salts ai present, the surface tension is reduced and the particles sink. In normal urines they remain floating.

See also: **bile acids and salts**

HCG STIMULATION TEST

A test of testicular function based on the fact that human chorionic gonadotrophin (HCG) resembles LH in its actions and stimulates the Leydig cells of the testis to secrete testosterone. It consists of injecting HCG and then measuring plasma testosterone in samples collected over several days. Patients with primary testicular failure fail to respond.

HEAT STABLE ALKALINE PHOSPHATASE

Synonym for placental alkaline phosphatase.

See also: **alkaline phosphatase**

HEAVY CHAIN DISEASES

A rare group of diseases in which there is an excess production of a protein which can be identified as part of the heavy chain of the immunoglobulin molecule. A paraprotein can be found in the blood in many of these cases. Five types of disease are theoretically possible but only three have been reported. Heavy chains of the γ type are produced in a generalized lymphoma (Franklin's disease). Heavy chains of the μ type are produced in a chronic lymphatic leukaemia, while heavy chains of the α type are found in an intestinal lymphoma.

Further reading: Hobbs, J.R. (1971). Immunoglobulins in clinical chemistry. In Bodansky, O. and Latner, A.L. (eds.) *Advances in Clinical Chemistry*. Vol. 14, p. 220. (New York: Academic Press)

HEMASTIX

A reagent stick manufactured by Ames for the detection of blood in urine. It is based on the ability of haemoglobin to act as a peroxidase, resulting in the oxidation of a dye to give a blue colour.

Further reading: Kutter, D. (1977). *Rapid Clinical Diagnostic Tests*. (Munich–Vienna–Baltimore: Urban and Schwarzenberg)

HEMATEST

A reagent kit manufactured by Ames for the detection of blood in faeces or urine. It is based on the peroxidase reaction.

Further reading: Kutter, D. (1977). *Rapid Clinical Diagnostic Tests*. (Munich–Vienna–Baltimore: Urban and Schwarzenberg)

HEMOCCULT TEST

A reagent kit manufactured by Smith, Kline and French Laboratories for the detection of occult blood. It is based on the peroxidase–guaiac reaction.

Further reading: Kutter, D. (1977) *Rapid Clinical Diagnostic Tests*. (Munich–Vienna–Baltimore: Urban and Schwarzenberg)

HENDERSON-HASSELBALCH EQUATION

This relates the pH of a solution to the concentration of base and acid as follows:

$$\text{pH} = \text{p}K_1 + \log \frac{[\text{base}]}{[\text{acid}]}$$

In the case of the plasma bicarbonate buffering system this can be expressed as:

$$\text{pH} = 6.10 + \log \frac{[\text{plasma bicarbonate}]}{0.03 \times P\text{CO}_2}$$

HEPARIN

A polysaccharide which acts as an anticoagulant by antagonizing the action of thrombin. It is used extensively as an anticoagulant in the collection of blood specimens, one of the reasons being that it does not cause a redistribution of water between plasma and red cells.

HEPATITIS

Inflammation of the liver. Acute hepatitis may be due to viruses or toxins. Chronic forms of hepatitis can also occur and some of these may have an autoimmune basis. The biochemical features of hepatitis are those of liver cell damage with high bilirubin and aminotransferase levels in the serum.

Further reading: Silk, D.B.A. and Williams, R. (1979). Acute liver failure. *Br. J. Hosp. Med.*, **22**, 437

HEPATITIS ASSOCIATED ANTIGEN

See: **Australia antigen**

HEPATOLENTICULAR DEGENERATION

See: **Wilson's disease**

HEPTABARBITONE

An intermediate acting barbiturate.

See: **barbiturates**

HEREDITARY ANGIONEUROTIC OEDEMA

A genetic condition characterized by a deficiency of C1 esterase inhibitor. This is an α_2-globulin which inhibits the action of C1 the first complement component, and in this way prevents possible damaging effects of the complement system. Patients affected with the condition suffer from oedematous swellings in various parts of the body, particularly the eyes and mouth.

Further reading: Ruddy, S. and Austen, R.F. (1978). Inherited abnormalities of the complement system. In Stanbury, J.B., Wyngaarden, J.B. and Fredrickson, D.S. (eds.) *The Metabolic Basis of Inherited Disease*. 4th Edn., p. 1737. (New York: McGraw-Hill)

HEREDITARY COPROPORPHYRIA

A type of porphyria which resembles acute intermittent porphyria in that certain drugs can precipitate acute attacks. It is characterized by large amounts of coproporphyrin in the faeces.

See also: **acute intermittent porphyria**

HEREDITARY FRUCTOSE INTOLERANCE

An inborn error of metabolism in which there are metabolic disturbances, especially hypoglycaemia, after fructose ingestion.

It is due to a deficiency of the enzyme fructose-1-phosphate aldolase. Fructosuria is a feature of the condition. In children, prolonged fructose ingestion leads to failure to thrive, hepatomegaly with jaundice, and vomiting.

Further reading: Froesch, E.R. (1978). Essential fructosuria, hereditary fructose intolerance and fructose-1,6-diphosphatase deficiency. In Stanbury, J.B., Wyngaarden, J.B. and Fredrickson, D.S. (eds.) *The Metabolic Basis of Inherited Disease*. 4th Ed., p. 121. (New York: McGraw-Hill)

HEREDITARY OROTIC ACIDURIA

A rare inborn error of pyrimidine metabolism in which there is an excessive excretion of orotic acid in the urine due to deficiencies of the enzymes involved in its metabolism. Affected individuals suffer from failure of normal growth and development. They also have a megaloblastic anaemia.

Further reading: Kelley, W.N. and Smith, L.H. Jr. (1978). Hereditary orotic aciduria. In Stanbury, J.B., Wyngaarden, J.B. and Fredrickson, D.S. (eds.) *The Metabolic Basis of Inherited Disease*. 4th Edn., p. 1045. (New York: McGraw-Hill)

HERS' DISEASE

Type VI glycogen storage disease.

See also: **glycogen storage diseases**

HEXOBARBITONE

An intermediate acting barbiturate.

See: **barbiturates**

HEXOSE-1-PHOSPHATE URIDYLTRANSFERASE (GALACTOSE-1-PHOSPHATE URIDYLTRANSFERASE)

The enzyme which is deficient in the most common form of galactosaemia.

See also: **galactosaemia**

5-HIAA

See: **5-hydroxyindoleacetic acid**

HIBITANE

A disinfectant which can be used as a urine preservative when for instance glucose is to be estimated.

HIGH DENSITY LIPOPROTEIN (HDL, α-LIPOPROTEIN)

A plasma lipoprotein fraction that contains mostly cholesterol and phospholipids. Recent work has suggested that the risk of coronary heart disease may be inversely related to the level of HDL cholesterol, i.e.. those with the highest level have the lowest risk. It has been suggested that HDL facilitates cholesterol removal from cells and carries it back to the liver for excretion. By this means the total body cholesterol is lowered.

Further reading: Editorial (1978). High density lipoprotein and atheroma. *Lancet,* **2**, 1291

HIGH PRESSURE LIQUID CHROMATOGRAPHY (HPLC)

A chromatographic technique in which the mobile phase is a liquid which is pumped at a high pressure over a densely packed stationary phase in a glass or stainless steel column. It enables high resolutions to be attained. Various types of chromatographic sparations can be used, including gel permeation chromatography, ion-exchange chromatography, adsorption (liquid–solid) chromatography and partition (liquid–liquid) chromatography.

A variety of detectors can be used for detecting material eluted from the column, e.g. UV, fluorimetric, polarographic, electron capture and flame ionization detectors. Since many of these detectors are non-destructive, the compounds eluted from the column can be used for further investigation, for instance by mass spectrometry.

HPLC has been used in the separation of many groups of compounds including steroids, porphyrins, drugs, proteins and lipids.

Further reading: Dixon, P.F., Stoll, M.S. and Lim, C.K. (1976). High pressure liquid chromatography in clinical chemistry: a review. *Ann. Clin. Biochem.,* **13**, 409

HIPPURIC ACID TEST

A test for investigating the conjugating capacity of the liver. It consists of administration of sodium benzoate to the patient followed by measurement of the urinary excretion of hippuric acid, a conjugate of benzoate and glycine. This test is now rarely performed.

HISTALOG STIMULATION TEST

A gastric function test based on the ability of Histalog (ametazole hydrochloride), a synthetic analogue of histamine, to stimulate acid secretion by parietal cells of the stomach. It is claimed to have fewer side-effects than histamine itself.

HISTAMINE STIMULATION TEST

A gastric function test based on the ability of histamine to stimulate the parietal cells of the stomach to produce hydrochloric acid. However histamine may have undesirable side effects and so 'Histalog', a synthetic analogue of histamine, may be used instead as this has fewer side effects.

HISTIDINAEMIA

A rare inborn error of metabolism characterized by increased blood histidine levels. It is due to a deficiency of the enzyme histidase. Histidine and imidazole are excreted in the urine. Mental retardation is found in many of the subjects. The mode of inheritance appears to be that of an autosomal recessive.

Further reading: La Du, B.N. (1978). Histidinaemia. In Stanbury, J.B., Wyngaarden, J.B. and Fredrickson, D.S. (eds.) *The Metabolic Basis of Inherited Disease*. 4th Ed., p. 317. (New York: McGraw-Hill)

HISTIDINE LOADING TEST (FIGLU TEST)

An *in vivo* test for the investigation of suspected folic acid deficiency. The normal metabolism of histidine contains a step in which formiminoglutamic acid (FIGLU) is converted to glutamate by an enzyme which uses folate as a cofactor. In patients with folate deficiency, administration of oral histidine results in a greater than normal urinary excretion of FIGLU.

This test is now rarely performed as there are relatively simple methods available for the direct determination of serum folate.

See also: **folic acid**

HOLLANDER TEST

A test for the completeness of vagotomy. Hypoglycaemia can result in stimulation of the vagus nerve and the subsequent secretion of gastric acid. A treatment for peptic ulcers consists of an operation in which the vagus nerve is cut. The completeness of the vagotomy can be ascertained by giving insulin to induce hypoglycaemia and then aspirating gastric fluid specimens in which the acid content is measured. Failure of the stomach to secrete acid suggests the vagotomy is complete.

HOMOCYSTINURIA

An inborn error of metabolism in which homocystine is excreted in the urine due to a deficiency of cystathionine synthase, the enzyme which catalyses the formation of cystathionine from homocysteine and serine. Homocysteine is a sulphur containing amino acid and can be detected in the urine by tests for this type of amino acid. Among the symptoms of the condition are minor congenital abnormalities, mental retardation and dislocated lenses.

Further reading: Mudd, S.H. and Levy, H.L. (1978). Disorders of transsulphuration. In Stanbury, J.B., Wyngaarden, J.B. and Fredrickson, D.S. (eds.) *The Metabolic Basis of Inherited Disease*. 4th Ed., p. 458. (New York: McGraw-Hill)

HOMOGENTISIC ACID

An organic acid found in the urine of patients with alkaptonuria, a hereditary condition in which there is a deficiency of homogentisic acid oxidase, one of the enzymes of tyrosine metabolism. As a result, homogentisic acid accumulates in the blood, tissues and urine. Urines containing homogentisic acid turn dark on standing due to oxidation and polymerization of the compound to the black pigment alkapton.

Detection of homogentisic acid in urine

Homogentisic acid can be detected by tests based on the fact that it is a powerful reducing agent. These tests include:

(1) The reduction of ammoniacal silver nitrate to give a black precipitate.

(2) The reduction of ferric chloride solution to give a transient blue colour.

(3) The reduction of Benedict's reagent.

See also: **alkaptonuria**

HOMOGENTISIC ACID OXIDASE

An enzyme involved in the breakdown of phenylalanine and tyrosine. A deficiency occurs in the inborn error of metabolism, alkaptonuria.

See also: **alkaptonuria**

HOMOVANNILIC ACID (HVA)

A metabolite of dopamine which is sometimes measured for the diagnosis of neuroblastoma. Gas chromatography, fluorimetry and colorimetry have been used for its measurement.

See also: **catecholamines**

HOPKINS-COLE REACTION

A test for tryptophan based on its reaction with glyoxylic acid to produce a purple colour. It is used in the measurement of total plasma globulins since these, unlike albumin, contain tryptophan.

HORMONES

Substances produced in small amounts in one part of the body and exerting their effects in another part of the body. Among the chemical classes of hormone are amines, steroid hormones, and polypeptide hormones.

HUMAN CHORIONIC GONADOTROPHIN (HCG)

A glycoprotein hormone synthesized by trophoblast cells. The detection of HCG in the urine is used as a pregnancy test. It is excreted in increased amounts in patients with a hydatidiform mole and in tumours of the testis or ovary. It was previously measured by biological assay but is now usually measured by immunochemical techniques. Immunochemical measurements of the β subunit of HCG can also be made. These are supposed to have a greater specificity than for intact HCG since the cross-reaction with LH is greatly reduced.

Further reading: Editorial (1977). Ubiquitous HCG. *Lancet*, **2**, 1116

HUMAN GROWTH HORMONE (HGH, SOMATOTROPHIN)

A polypeptide hormone secreted by the anterior pituitary. Growth hormone secretion is suppressed by hyperglycaemia and stimulated by hypoglycaemia. Two hypothalamic hormones are involved in its release, growth hormone releasing hormone (GHRH) which stimulates its release, and growth hormone release inhibitory hormone (GHR-IH, somatostatin) which, as its name suggests, inhibits its release. Growth hormone has a central role in the promotion of growth. It is usually measured by radioimmunoassay. Among its actions are:

(1) Stimulation of skeletal growth, an action which is mediated by a small peptide, somatomedin, synthesized in the liver.

(2) Stimulation of protein synthesis.

(3) Opposition of the actions of insulin, thus tending to raise the blood sugar levels.

(4) Stimulation of lipolysis.

Increased serum growth hormone levels

(1) In gigantism. Excess growth hormone secretion in childhood leads to this condition.

(2) In acromegaly.

Decreased serum growth hormone levels

In hypothalamic or pituitary hypofunction, deficient growth hormone secretion leads to dwarfism.

Further reading: Hall, R., Anderson, J., Smart, G.A. and Besser, M. (1974). *Fundamentals of Clinical Endocrinology*. 2nd Edn. (London: Pitman Medical Publishing Co.)

HUMAN MENOPAUSAL GONADOTROPHIN

These are gonadotrophins isolated from the urine of post-menopausal women. They are used therapeutically to induce ovulation.

HUMAN PLACENTAL LACTOGEN (HPL, HUMAN CHORIONIC SOMATOMAMMOTROPHIN)

A polypeptide hormone secreted in large amounts by the placenta (1–3 g/day near term). Maternal serum levels rise continually throughout pregnancy until about the 36th week. Its exact role during pregnancy has not yet been elucidated. HPL has a half-life of about 25 minutes and therefore plasma levels rapidly reflect changes in its production. A fall in the serum level during pregnancy suggests placental dysfunction and this enables early obstetric action to be taken. HPL is usually measured by radioimmunoassay.

Further reading: Wilde, C.E. and Oakey, R.E. (1975). Scientific Review No. 3. Biochemical tests for the assessment of feto-placental function. *Ann. Clin. Biochem.*, **12**, 83

HUNTER'S SYNDROME

A type of mucopolysaccharidosis.

See: **mucopolysaccharidoses**

HUNTER'S TEST

A test for the detection of bilirubin in urine based on a diazotization reaction

HURLER'S SYNDROME

A type of mucopolysaccharidosis.

See: **mucopolysaccharidoses**

HYDATIDIFORM MOLE

A tumour like growth in parts of the placenta. Human chorionic gonadotrophin excretion is associated with this condition.

HYDROCORTISONE

See: **cortisol**

HYDROCORTISONE SUPPRESSION TEST (STEROID SUPPRESSION TEST)

A test for the differential diagnosis of hypercalcaemia. Hydrocortisone or cortisone administration causes a fall in the serum calcium levels in all cases of hypercalcaemia except primary or tertiary hyperparathyroidism. The reasons for this are not well understood.

See also: **calcium**

HYDROGEN BREATH TEST

A test which can be used in the diagnosis of lactose malabsorption, based on the fact that hydrogen is evolved when malabsorbed carbohydrate is fermented by colonic bacteria. A proportion of the gas diffuses into the circulation and is carried to the lungs where it is exhaled. Gas chromatography can be used to measure the amount of hydrogen in the expired air.

Further reading: Editorial. (1975). Breath tests in gastroenterology. *Lancet*, 2, 163

HYDROGEN ION CONCENTRATION

See: **pH**

11-β-HYDROXYAETIOCHOLANOLONE

A urinary metabolite of cortisol.

β-HYDROXYBUTYRATE

A ketone body formed as a result of fatty acid oxidation.

See also: **ketone bodies**

α-HYDROXYBUTYRATE DEHYDROGENASE (α-HBD, SERUM HYDROXYBUTYRATE DEHYDROGENASE, SHBD)

The heart isoenzyme of lactate dehydrogenase which can catalyse the oxidation of hydroxybutyrate to oxobutyrate. Its measurement therefore in serum can be used as an indication of the level of the heart isoenzyme present.

Increases in serum levels

Increases are found after myocardial infarction reaching a peak level at about 2–3 days after the event. Raised levels have also been found in megaloblastic anaemias, acute leukaemias and renal damage.

Measurement

In most methods for its estimation, oxobutyrate is used as the substrate. Its reduction to hydroxybutyrate is accompanied by the oxidation of NADH to NAD which can be followed spectrophotometrically. Some colorimetric methods are also available. For instance, the oxobutyrate remaining after the reaction can be determined colorimetrically as the dinitrophenylhydrazone derivative.

Further reading: Wilkinson, J.H. (1976). *The Principles and Practice of Diagnostic Enzymology*. (London: Edward Arnold)

25-HYDROXYCHOLECALCIFEROL

A derivative of vitamin D formed by the hydroxylation of cholecalciferol in the liver. It is further hydroxylated by the kidney to form 1,25-dihydroxycholecalciferol, the biologically active form of the vitamin.

See also: **vitamin D**

11-HYDROXYCORTICOSTEROIDS

A collective term for a group of corticosteroids having a hydroxyl group at the 11-position, which can be estimated fluometrically in plasma and urine. Since the main steroid measured is cortisol, with a minor contribution from corticosterone, the term for most clinical purposes is synonymous with cortisol. Increases in plasma and urine 11-hydroxycorticosteroid levels are found in Cushing's syndrome whereas decreased levels are found in Addison's disease.

Measurement

The fluorimetric technique for their estimation consists of

extracting the steroids into dichloromethane, adding sulphuric acid–ethanol reagent and measuring the fluorescence in this solvent.

See also: **cortisol**

17-HYDROXYCORTICOSTEROIDS (17-HYDROXYSTEROIDS)

This is a collective term for a group of corticosteroids which includes cortisol, cortisone, 11-deoxycortisol and their tetrahydro derivatives which are measured in plasma and urine by the Porter–Silber reaction. The term 17-hydroxysteroids is often used synonymously with 17-oxogenic steroids and to most intents and purposes they mean the same. However there is a subtle difference between the two in that 17-oxogenic steroids (determined by the Zimmermann reaction after reduction and oxidation) measures, in addition to the compounds mentioned above, cortol and cortolone.

Decreased urinary excretion of 17-hydroxycorticosteroids occurs in Addison's disease while increases are found in Cushing's syndrome.

See also: **Porter-Silber reaction**

5-HYDROXYINDOLEACETIC ACID

A metabolite of the hormone, serotonin (5-hydroxytryptamine). It is excreted in excess amounts in the urine of patients with carcinoid syndrome, a condition in which there is excessive synthesis of serotonin.

It can be detected in urine by the formation of a purple colour with 1-nitroso-2-naphthol and nitrous acid. This reaction can also be used for its quantitation.

See also: **carcinoid syndrome**

4-HYDROXY, 3-METHOXYMANDELIC ACID (HMMA) (VANILLYL-MANDELIC ACID, VMA)

An end product of catecholamine catabolism. It is excreted in the urine in large amounts in patients with catecholamine secreting tumours such as phaeochromocytoma and neuroblastoma.

Measurement

(1) HMMA can be measured by its reaction with diazotized *p*-nitroaniline to give a coloured compound. Certain fruits, drugs and vanillin containing foods must be excluded from the diet prior to the start of the urine collection, as these may give falsely elevated values.

(2) However no dietary restrictions are usually necessary if the HMMA is determined by the method of Pisano. This consists of extracting HMMA from acidified urine into ethyl acetate and then back-extracting into potassium carbonate solution. To this is added metaperiodate which oxidizes HMMA to vanillin and this is extracted into toluene and measured spectrophotometrically at 360 nm.

See also: **catecholamines**

17-HYDROXYPREGNANOLONE

This is a metabolite of 17-hydroxyprogesterone, an intermediate in cortisol biosynthesis. It is found in increased amounts in the urine, together with its metabolite pregnanetriol, in some forms of congenital adrenal hyperplasia.

See also: **congenital adrenal hyperplasia**

17-HYDROXYPROGESTERONE

An intermediate in the formation of cortisol. In some forms of congenital adrenal hyperplasia it accumulates in increased amounts and can be measured in plasma. It is metabolized to 17-hydroxypregnanolone and then to pregnanetriol which can be measured in the urine in order to diagnose the condition.

See also: **congenital adrenal hyperplasia**

HYDROXYPROLINAEMIA

An inborn error of metabolism in which elevated hydroxyproline levels are found in the blood due to a deficiency in the enzyme, hydroxyproline oxidase. Mental subnormality is a feature.

Further reading: Scriver, C.R. (1978). Disorders of proline and hydroxyproline metabolism. In Stanbury, J.B.

Wyngaarden, J.B. and Fredrickson, D.S. (eds.) *The Metabolic Basis of Inherited Disease*. 4th Edn., p. 336. (New York: McGraw-Hill)

HYDROXYPROLINE

An imino acid which can be measured in two types of clinical condition:

(1) Inborn errors.

(a) Increased plasma hydroxyproline levels occur in the inborn error of metabolism, hydroxyprolinaemia, due to a deficiency in the enzyme hydroxyproline oxidase.

(b) Increased urinary excretion of hydroxyproline, together with proline and glycine, occurs in iminoglycinuria when there is defective renal tubular transport of these compounds. Two types of iminoglycinuria have been described, a mild form and a more severe form where the symptoms include mental retardation and epilepsy.

(2) Bone diseases. Bone contains about 40% of the total body collagen and this is the only protein that contains significant amounts of hydroxyproline. Providing the patient is on a hydroxyproline-free diet, quantitative estimations of urinary hydroxyproline give an indication of the turnover of bone matrix. High urinary hydroxyproline levels are therefore found in bone diseases where there is a high rate of matrix turnover, such as Paget's disease of bone or metastatic bone disease. Measurement of urinary hydroxyproline can be used to monitor the progress of patients with Paget's disease who are receiving treatment with calcitonin.

Measurement

Hydroxyproline in urine can be measured by its oxidation to pyrrole (usually by chloramine T) followed by its reaction with Ehrlich's reagent to form a red compound.

Further reading: LeRoy, E.C. (1967). The technique and significance of hydroxyproline measurement in man. In Bodansky, O. and Stewart, C.P. (Eds.) *Advances in Clinical Chemistry*. Vol. 11, p. 213. (New York: Academic Press)

5-HYDROXYTRYPTAMINE (SEROTONIN)

A hormone formed from tryptophan as follows:

tryptophan —hydroxylation→ 5-hydroxytryptophan

5-hydroxytryptophan —decarboxylation→ 5-hydroxytryptamine (serotonin)

It is synthesized by argentaffin cells of the alimentary tract. It stimulates smooth muscle contraction and is a powerful vasoconstrictor. Tumours of argentaffin cells cause the condition known as carcinoid syndrome, some of the symptoms of which can be related to the excess secretion of 5-hydroxytryptamine.

5-Hydroxytryptamine is usually measured as its urinary metabolite 5-hydroxyindoleacetic acid.

See also: **carcinoid syndrome**

5-HYDROXYTRYPTOPHAN

An intermediate in the conversion of tryptophan to 5-hydroxy tryptamine (serotonin). In some cases of carcinoid syndrome it i excreted in large amounts in the urine, even though the 5 hydroxyindoleacetic acid excretion may be normal. This i thought to be because the cells lack the decarboxylase whicl converts 5-hydroxytryptophan to 5-hydroxytryptamine.

See also: **carcinoid syndrome**

HYPER . . . AEMIA

If not listed below, see under relevant substance, e.g. for hyper glycaemia, see **glucose.**

HYPERCHLORHYDRIA

The secretion of excess gastric acid. This can result in gastric c duodenal ulceration. The excess secretion can be neurogenic i origin, or it can also occur in Zollinger–Ellison syndrome, whe large amounts of gastrin are produced.

HYPERLIPOPROTEINAEMIAS

Hyperlipoproteinaemias can arise by a number of different mechanisms:

(1) As a hereditary condition, i.e. familial hyperlipoproteinaemias.

(2) As a result of a high dietary intake of calories, saturated fats or alcohol.

(3) Secondary to another condition, e.g. hypothyroidism or diabetes mellitus.

(4) Stress induced.

Classification

The hyperlipoproteinaemias can be classified in a number of ways.

The Strisower classification is based on the behaviour of the lipoproteins in the ultracentrifuge. The Thorpe and Stone classification classifies the lipoproteins on the basis of their size. The Fredrickson classification characterizes the lipoproteins on the basis of their electrophoretic behaviour and is described here:

(1) Type I. This is characterized by the persistent presence of chylomicrons in the plasma.

(2) Type IIa. This is characterized by an increase in plasma β lipoproteins.

(3) Type IIb. This is characterized by an increase in plasma β and pre-β-lipoproteins.

(4) Type III. This is characterized by the presence of a 'broad β band' extending from the β to the pre-β region.

(5) Type IV. This is characterized by an increase in plasma pre-β-lipoproteins.

(6) Type V. In this type, there is an increase in both chylomicrons and pre-β-lipoproteins.
The commonest types are IIa and IV.

Treatment

Dietary restrictions and drugs such as cholestyramine or clofibrate are used in the treatment of hyperlipoproteinaemias.

Further reading: Chait, A. (1978). Hyperlipoproteinaemia - an approach to diagnosis and classification. In Alberti, K.G.M.M. (ed.) *Recent Advances in Clinical Biochemistry*. Vol. 1, p. 73. (Edinburgh, London and New York: Churchill-Livingstone)

HYPERPARATHYROIDISM

These are diseases in which there is a high circulating parathyroid hormone (PTH) level.

Primary hyperparathyroidism

This occurs when there is a parathyroid adenoma or ectopic PTH secretion from a tumour. It can cause hypercalcaemia with its resultant clinical symptoms.

Secondary hyperparathyroidism

This is when the parathyroids are stimulated to produce PTH as a result of a hypocalcaemic condition, e.g. calcium or vitamin D deficiency.

Tertiary hyperparathyroidism

This is when the parathyroid gland has been under prolonged stimulation as a result of a long standing hypocalcaemic condition, with the result that PTH hypersecretion has become autonomous, even when the original cause of the hypocalcaemia has been corrected. Apart from the original hypocalcaemia the condition is similar to primary hyperparathyroidism.

See also: **calcium, parathyroid hormone**

Further reading: Hall, R., Anderson, J., Smart, G.A. and Besser, M. (1974). *Fundamentals of Clinical Endocrinology*. 2nd Edn. (London: Pitman Medical Publishing Co.,)

HYPERPROLINAEMIA

See: **proline**

HYPERTHYROIDISM (THYROTOXICOSIS)

The condition that results from excess thyroid hormone secre tion. These accelerate many of the processes occurring in th body and the clinical features reflect this. They include weigh loss, sweating, tachycardia, tremor, diarrhoea and emotional disturbances. The commonest causes of hyperthyroidism are:

(1) Graves' disease. In this autoimmune condition the thyroid gland is stimulated by an immunoglobulin called long-acting thyroid stimulator (LATS).

(2) Autonomous secretion of thyroid hormones by toxic nodules or adenomas of the thyroid gland.

Biochemical tests used in the investigation of hyperthyroidism

(1) The serum thyroxine is usually raised. Occasionally it is normal, but the serum tri-iodothyronine (T3) is raised ('T3 toxicosis'). This may represent an early phase of hyperthyroidism.

(2) TSH levels are low.

(3) The result of the radioactive iodine neck uptake test may be high.

(4) T3 suppression test. The neck uptake is repeated after previously having given T3 which normally suppresses TSH production. Lack of suppression suggests hyperthyroidism.

See also: **thyroxine**

Further reading: Hall, R., Anderson, J., Smart, G.A. and Besser, M. (1974). *Fundamentals of Clinical Endocrinology*. 2nd Edn. (London: Pitman Medical Publishing Co.)

Editorial. (1979). Four controversies about the cause of hyperthyroidism. *Lancet*, **2**, 78

HYPO . . . AEMIA

If not listed below see under relevant substance, e.g. for hypocalcaemia, see under **calcium**.

HYPOGONADISM

This presents in children as delayed puberty or in adults as infertility. It can be a primary condition due to gonadal dysfunction or secondary to hypothalamic or pituitary disorders.

Biochemical tests used in the investigation of hypogonadism

(1) Testosterone (in males) or oestrogen (in females) levels are low.

(2) Gonadotrophin levels are low in secondary hypogonadism but raised in primary hypogonadism.

(3) Clomiphene stimulation test. An impaired response is found when there is pituitary dysfunction.

(4) LH/FSH releasing hormone test. In some cases of secondary hypogonadism an impaired response occurs.

(5) HCG stimulation test (in males). In primary hypogonadism, the plasma testosterone level does not rise.

Further reading: Hall, R., Anderson, J., Smart, G.A. and Besser, M. (1974). *Fundamentals of Clinical Endocrinology*. 2nd Edn. (London: Pitman Medical Publishing Co.)

HYPOPARATHYROIDISM

A condition in which there is a low circulating parathyroid hormone (PTH) level.

Primary hypoparathyroidism can be due to atrophy of the parathyroid gland, possibly by an autoimmune process. Another cause is removal of the glands during thyroidectomy. Primary hypothyroidism can result in hypocalcaemia with its associated clinical symptoms.

Secondary hypoparathyroidism results from suppression of PTH secretion in diseases in which there is hypercalcaemia, e.g. sarcoidosis, thyrotoxicosis and malignant bone disease.

(N.B. In pseudohypoparathyroidism, PTH is secreted normally but the renal tubules cannot respond to it.)

See also: **calcium, parathyroid hormone**

Further reading: Hall, R., Anderson, J., Smart, G.A. and Besser, M. (1974). *Fundamentals of Clinical Endocrinology*. 2nd Edn. (London: Pitman Medical Publishing Co.)

HYPOPITUITARISM

This can result from disease of the hypothalamus, resulting in failure to secrete the releasing factors, or from disease of the pituitary itself. The features of the condition reflect the deficiencies of the pituitary hormone, i.e. dwarfism due to HGH deficiency, hypothyroidism from TSH deficiency, adrenocortical hypofunction from ACTH deficiency and disturbances in the sexual characteristics from gonadotrophin deficiency. The condition can be diagnosed by measurement of the pituitary hormones in the blood.

Further reading: Hall, R., Anderson, J., Smart, G.A. and Besser, M. (1974). *Fundamentals of Clinical Endocrinology*. 2nd Edn. (London: Pitman Medical Publishing Co.)

HYPOTHALAMIC RELEASING AND INHIBITING FACTORS

Small polypeptide hormones released by the hypothalamus which regulate the release of a specific pituitary hormone. They include:

(1) Growth hormone releasing factor (GHRF).

(2) Thyroid stimulating hormone releasing factor (TSH-RF), sometimes known as thyrotrophin-releasing factor (TRF).

(3) Luteinizing hormone and follicle-stimulating hormone releasing factor (LH/FSH-RF).

(4) Corticotrophin releasing factor (CRF)

These four factors stimulate the release of the appropriate pituitary hormone. In contrast, hypothalamic inhibitory factors have been discovered, which inhibit the release of a pituitary hormone. These include:

(1) Melanocyte inhibitory factor (MIF) which inhibits melanocyte stimulating hormone (MSH) release.

(2) Prolactin-release inhibiting factor (PIF).

(3) Growth hormone release inhibitory hormone (GHR-IH, somatostatin).

Further reading: Malarkey, W.B. (1976). Recently discovered hypothalamic–pituitary hormones. *Clin. Chem.* **22**, 5

Hall, R. and Gomez-Pan, A. (1976). The hypothalamic regulatory hormones and their clinical applications. In Bodansky, O. and Latner, A.L. (eds.) *Advances in Clinical Chemistry*. Vol. 18, p. 174. (New York: Academic Press)

Supplement. Hypothalamic and pituitary hormones. *J. Clin. Pathol.* April 1979

HYPOTHYROIDISM

The condition that results from inadequate secretion of thyroid hormones. The clinical features include weight gain, dry skin, loss of hair, a hoarse voice and thickening of the subcutaneous tissues (myxoedema). If the disease goes undetected, a myxoedema coma may develop.

Primary hypothyroidism occurs when there is destruction of the thyroid gland by circulating thyroid autoantibodies. Secondary hypothyroidism occurs when there is TSH deficiency due to hypothalamic or pituitary disease.

Biochemical tests used in the investigation of hypothyroidism

(1) The serum thyroxine level is low.

(2) TSH is increased in primary hypothyroidism and decreased in secondary hypothyroidism.

(3) A TRH stimulation test may be performed which involves injection of TRH followed by measurement of TSH in the serum. Lack of any increase in TSH levels suggests pituitary hypofunction whereas a normal response suggests hypothalamic disease.

(4) Radioactive iodine uptake test. Lower than normal neck uptakes suggest hypothyroidism. This can followed by:

(5) TSH stimulation test. This involves injecting TSH and measuring the radioactive iodine neck uptake the following day. Lack of any increase in uptake is usually indicative of primary hypothyroidism. Some response usually occurs in secondary hypothyroidism.

(6) The detection of circulating thyroid autoantibodies can be used for the diagnosis of primary hypothyroidism.

See also: **thyroxine**

Further reading: Hall, R., Anderson, J., Smart, G.A. and Besser, M. (1974). *Fundamentals of Clinical Endocrinology.* 2nd Edn. (London: Pitman Medical Publishing Co.)

HYPOXANTHINE-GUANINE PHOSPHORIBOSYL TRANSFERASE

An enzyme involved in the resynthesis of purine nucleotides from xanthine, hypoxanthine and guanine. A deficiency of the enzyme occurs in the inborn error of metabolism Lesch–Nyhan syndrome (juvenile hyperuricaemia).

See also: **Lesch–Nyhan syndrome**

I

ICTERIC INDEX

An old method of measuring the degree of jaundice which consists of assessing the intensity of the yellow colour of the serum. The serum is diluted until it matches the colour of a 1 in 10 000 solution of potassium dichromate. The dilution factor is termed the icteric index.

ICTOSTIX

A reagent stick manufactured by Ames for the detection of bilirubin in urine. It is based on the reaction of bilirubin with diazotized 2,4-dichloroaniline.

Further reading: Kutter, D. (1977). *Rapid Clinical Diagnostic Tests.* (Munich–Vienna–Baltimore: Urban and Schwarzenberg)

ICTOTEST

A tablet test manufactured by Ames for the detection of bilirubin in urine. It is based on the reaction of bilirubin with *p*-nitrobenzene diazonium *p*-toluene sulphonate to produce a purple colour.

Further reading: Kutter, D. (1977). *Rapid Clinical Diagnostic Tests.* (Munich–Vienna–Baltimore: Urban and Schwarzenberg)

IgG, IgA, IgM, IgD, IgE

See: **immunoglobulins**

IMINOGLYCINURIA

Familial iminoglycinuria is an inborn error in which there is increased excretion of proline, hydroxyproline and glycine in the urine. It is due to defective renal tubular transport of these compounds. It has an autosomal recessive mode of inheritance. One form of the condition appears to be relatively harmless, but a more severe form can occur in which there is mental retardation and epilepsy.

Further reading: Scriver, C.R. (1978). Disorders of proline and hydroxyproline metabolism. In Stanbury, J.B., Wyngaarden, J.B. and Fredrickson, D.S. (eds.) *The Metabolic Basis of Inherited Disease*. 4th Edn., p. 336. (New York: McGraw-Hill)

IMIPRAMINE

A trycyclic antidepressant drug. It can be detected in urine by a number of screening tests based on its reaction with oxidizing agents to give a blue colour.

IMMUNE COMPLEX DISEASES

See: **collagen diseases**

IMMUNOCHEMICAL TECHNIQUES

Techniques for the detection or assay of various substances based on the reaction of those substances with specific antibodies (or *vice versa,* i.e. the detection and assay of antibodies using antigens). Such techniques include agglutination reactions, automated immune precipitation, complement fixation tests, crossed electrophoresis, counter electrophoresis, double diffusion, enzyme immunoassay, fluoroimmunoassay, haemagglutination, immunoelectrophoresis, immunofluoresence, radial immunodiffusion, spin immunoassay, immunofixation, immunoradiometric assay and radioimmunoassay. See separate entries for these subjects.

IMMUNOCYTOMA

A disease in which there is a proliferating clone of immunoglobulin-producing cells (plasma cells or lymphocytes). Myeloma, macroglobulinaemia and heavy chain diseases are examples of such diseases and they are often characterized by the presence of an abnormal protein (a paraprotein) in the blood.

See also: **heavy chain diseases, macroglobulinaemia, myeloma**

IMMUNODIFFUSION

The detection or estimation of antigens (or antibodies) by their precipitation in gels after having been allowed to diffuse towards antibody (or antigen).

See also: **double diffusion, radial immunodiffusion**

IMMUNOELECTROPHORESIS

A technique that is used for the identification of proteins. It consists initially of separation of the proteins of the serum, or other fluid, by electrophoresis. This is usually carried out in agarose, although other support materials can be used. After electrophoresis, antiserum is placed in a trough adjacent and parallel to the direction of electrophoresis and the proteins and antibodies are allowed to diffuse towards each other. Interaction between the antibodies and the protein antigen results in the precipitation of immune complexes within the gel, which takes the form of an arc or bow for each protein. Thus if human serum has been separated and polyvalent antiserum has been placed in the trough, a series of arcs is obtained, each arc resulting from the precipitation of a serum protein with its own specific antibody. Polyvalent antiserum or monospecific antisera (anti-IgG, anti-IgA etc.) can be placed in the troughs. In the case of mono-specific antiserum, usually only one precipitin bow results.

If a paraprotein is present, an abnormally shaped precipitation arc may be found with its corresponding antiserum. Thus a serum or urine paraprotein can be identified by electrophoresis of several portions of the sample, followed by the addition of specific antisera (anti-IgG, anti-IgA etc.) to each of the troughs. An abnormally shaped precipitation arc against for instance anti-IgA suggests that an IgA paraprotein is present.

Further reading: Grant, G.H. and Butt, W.R. (1970). Immunochemical methods in clinical chemistry. In Bodansky, O. and Stewart, C.P. (eds.) *Advances in Clinical Chemistry*. Vol. 13, p. 383. (New York: Academic Press).

Riches, P.G. (1979). Electrophoresis and immunoelectrophoresis. In Milford Ward, A. and Whicher, J.T. (eds.) *Immunochemistry in Clinical Laboratory Medicine*. p. 3. (Lancaster: MTP Press)

IMMUNOFIXATION

A technique for the identification of proteins in biological fluids. The proteins are first separated electrophoretically and cellulose acetate strips, each containing a specific antiserum, are placed over the electrophoretograms. After allowing the antibody–antigen reaction to take place, the unreacted proteins are washed away and the antibody–antigen complex visualized by staining. In this way, a protein can be identified by determining which antiserum it reacts against.

Further reading: Ritchie, R.F. and Smith, R. (1976). Immunofixation. I. General principles and application to agarose gel electrophoresis. *Clin. Chem.*, **22**, 497

IMMUNOFLUORESCENCE

A technique for the microscopic localization of antigens in or on various cells. In the direct staining technique, a fluorescent dye (e.g. fluoroscein) is linked to an antibody which is added to the tissue preparation. Sites where antibody has combined with antigen exhibit a brilliant fluorescence which can be seen microscopically.

An indirect staining technique is used for the detection of autoantibodies, in which a tissue extract is treated with the patients serum. The preparation is then treated with anti-human globulin linked to a fluorescent dye. This combines with bound autoantibodies on the tissue extract and the fluorescence can be observed microscopically.

Further reading: Roitt, I.M. (1977). *Essential Immunology*. 3rd Edn. (Oxford, London, Edinburgh, Melbourne: Blackwell Scientific Publications)

IMMUNOGEN

A substance which, when injected into animals, is capable of inducing the formation of antibodies. The term is often used synonymously with antigen, although, strictly speaking, the term antigen should refer to any substance which is capable of reacting with antibody (e.g. *in vitro*), without necessarily having the ability to stimulate antibody production *in vivo*.

IMMUNOGLOBULINS

Proteins which function as antibodies and thus are part of the body's defence mechanism. The basic immunoglobulin unit consists of two heavy chains and two light chains linked by disulphide bridges.

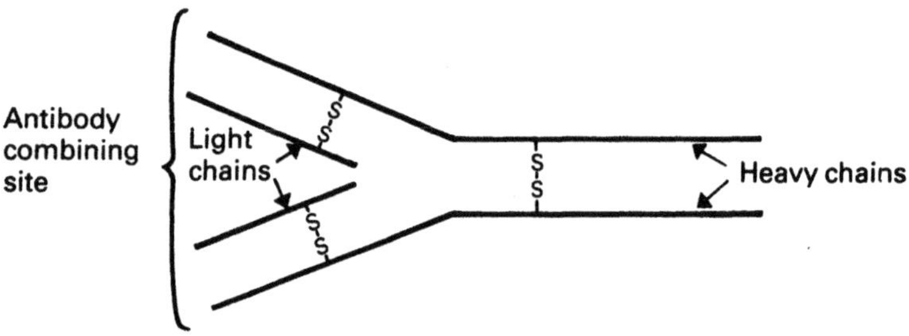

There are five types of heavy chain, $\gamma, \alpha, \mu, \delta$ and ϵ, and these give rise to the five classes of immunoglobulin, namely IgG, IgA, IgM, IgD and IgE. There are two types of light chain, kappa (κ) and lambda (λ).

Most immunoglobulins migrate electrophoretically in the γ region, although some have β, or even α-globulin mobility.

The different immunoglobulin classes (in order of their levels in blood)

(1) IgG. This appears to function in protecting body fluids. It is active against soluble antigens.

(2) IgA. This is synthesized beneath the mucosa of the respiratory and gastrointestinal tracts where its role is in the protection of body surfaces. It is present in secretory fluids in the form of a dimer (secretory IgA), two immunoglobulin units linked by a small protein (the secretory piece).

(3) IgM. This is a pentamer of five immunoglobulin units. It is active against particulate antigens such as microorganisms.

(4) IgD. The function of this immunoglobulin is not known.

(5) IgE. This is involved in allergic reactions. In the circulation it is bound to mast cells and basophils and, when it

combines with antigen, the cells release substances which cause the hypersensitivity reaction.

Increases in serum immunoglobulin levels

An increase in the serum immunoglobulin levels may be due to a polyclonal increase of an immunoglobulin class or several classes, or a monoclonal increase of one particular immunoglobulin.

Polyclonal increases. These result from the proliferation of many types of immunoglobulin-producing cells synthesizing a range of immunoglobulin molecules. Infections can result in generalized increases in all three of the major immunoglobulin classes (IgG, IgA and IgM). In some conditions a single immunoglobulin class is increased, e.g. IgG in some autoimmune diseases, IgA in disease of the gastrointestinal or respiratory tracts and liver cirrhosis, IgM in parastic infestations and primary biliary cirrhosis, and IgE in parasitic infections and allergic conditions.

Monoclonal increases. These result from the proliferation of a single type of immunoglobulin synthesizing cell, producing a single class and type of immunoglobulin, known as a paraprotein. Paraproteins are found in myeloma, Waldenström's macroglobulinaemia and heavy-chain diseases.

Decreases in serum immunoglobulin levels

Decreased immunoglobulin levels are found in conditions where there is deficient protein synthesis, such a malnutrition or malabsorption, or excessive protein loss, such as nephrotic syndrome. There are several rare congenital syndromes in which there is a complete or partial deficiency of one or more classes of immunoglobulins.

Measurement of serum immunoglobulin levels

The immunoglobulins present in the serum in the greatest amounts (IgG, IgA and IgM) can be measured by a variety of immunochemical tests using antiserum directed against one particular class of immunoglobulin (anti-IgG etc.). The techniques that can be used include radial immunodiffusion, automated immune precipitation and the Laurell 'rocket' technique. Because of its much smaller concentrations, IgE is usually measured by radioimmunoassay.

Further reading: Hobbs, J.R. (1971). Immunoglobulins in clinical chemistry. In Bodansky, O. and Latner, A.L. (eds.)

Advances in Clinical Chemistry. Vol. 14, p. 220. (New York: Academic Press)

IMMUNORADIOMETRIC ASSAY

A technique for the estimation of antigens in which they react with radioactively labelled antibody. By counting the antibody–antigen complex, a direct indication of the amount of antigen can be obtained.

INAPPROPRIATE HORMONE SECRETION

The secretion of a hormone under conditions when its secretion should normally have been stopped. Many causes of inappropriate hormone secretion are due to tumours but it can also occur in other conditions, e.g. infections.

INDICAN

A compound formed by the action of bacteria in the gut on dietary tryptophan. It is absorbed from the gut and subsequently excreted in the urine. Increased bacterial activity in the gut (e.g. in blood loop syndrome or inestinal stagnation) results in increased urinary indican excretion.

Detection and measurement

Indican can be detected in urine by its reaction with Ehrlich's reagent or by the Jaffé test in which a blue colour is produced in the presence of hydrochloric acid and calcium hypochlorite which can be extracted into chloroform. It can be quanititated by its reaction with Ehrlich's reagent, the chromogen being extracted into alkaline solution and estimated spectrophotometrically.

INDOCYANINE GREEN

A dye which can be used in a similar manner to bromsulphthalein to test hepatic excretory function.

INFRA-RED SPECTROSCOPY

Radiation in the infra-red region causes vibrational transitions in the bonds of molecules. The measurement of the infra-red radiation absorbed by a compound enables its infra-red ab sorption spectrum to be found. Every compound has its own characteristic absorption spectrum and thus unknown molecules, e.g. drugs, can be identified in this way.

Further reading: Broughton, P.M.G. and Dawson, J.B. (1972). Instrumentation in clinical chemistry. In Bodansky, O and Latner, A.L. (eds.) *Advances in Clinical Chemistry*. Vol. 15, p. 288. (New York: Academic Press)

INORGANIC PHOSPHATE

See: **phosphate**

INSULIN

A polypeptide hormone secreted by the β cells of the pancreatic islets of Langerhans. It is instrumental in lowering blood glucose levels. This occurs as a result of its action in promoting the entry of glucose into cells, by stimulating glycogenesis and inhibiting gluconeogenesis. It is also involved in lipid metabolism, inhibiting lipolysis and stimulating lipogenesis. Together with growth hormone, it stimulates protein synthesis.

Control of insulin secretion

(1) The most important physiological factor is probably the blood glucose level, the higher the level the more insulin is secreted.

(2) Some amino acids, particularly leucine and arginine, stimulate insulin secretion.

(3) Glucagon stimulates insulin secretion.

(4) Oral hypoglycaemic drugs, such as tolbutamide, are used in the treatment of diabetes mellitus as these stimulate insulin secretion.

(5) In contrast insulin release is inhibited by adrenaline.

Abnormal blood insulin levels

In diabetes mellitus the plasma level is generally low. In patients with islet cell tumours of the pancreas (insulinoma) plasma levels are high.

Although insulin can be measured by bioassay, it is usually measured by radioimmunoassay.

See also: **diabetes mellitus**

Further reading: Hall, R., Anderson, J., Smart, G.A. and Besser, M. (1974). *Fundamentals of Clinical Endocrinology*. 2nd Edn., (London: Pitman Medical Publishing Co.)

INSULIN ANTIBODIES

Antibodies, against insulin, produced by some patients who are receiving insulin treatment. They may be responsible for the unstable diabetic control in many of these patients. The antibodies can be measured in blood by immunoassay.

INSULIN-LIKE SUBSTANCE

A substance with insulin-type actions which may be produced in some tumours.

INSULINOMA

A tumour of the islet cells of the pancreas resulting in excess insulin secretion and therefore hypoglycaemia. The condition can be diagnosed by demonstrating inappropriately high levels of insulin in the blood. Several dynamic tests can also be used to diagnose the condition.

(1) Glucagon tolerance test. Glucagon stimulates insulin secretion. In patients with insulinoma, glucagon administration results in severe hypoglycaemia, as a result of excess insulin secretion.

(2) Leucine test. This amino acid, like glucagon, also stimulates insulin secretion. In patients with insulinoma severe hypoglycaemia results.

(3) Tolbutamide test. Tolbutamide is a drug which stimulates insulin secretion. In patients with insulinoma severe hypoglycaemia results after its intravenous injection.

The usual treatment of insulinoma is by removal of the gland.

Further reading: Hall, R., Anderson, J., Smart, G.A. and Besser, M. (1974). *Fundamentals of Clinical En-*

docrinology. 2nd Edn. (London: Pitman Medical Publishing Co.)

INSULIN STRESS TEST

An *in vivo* test for the assessment of growth hormone and ACTH secretion in response to the hypoglycaemia induced by insulin. It consists of insulin administration followed by the collection of blood samples for up to two hours. Glucose (to check that hypoglycaemia has in fact been induced), cortisol and growth hormone are estimated in the samples. Failure of the growth hormone levels to rise significantly is suggestive of hypothalamic or anterior pituitary hypofunction. Failure of the cortisol levels to rise significantly is indicative of a defect of the hypothalamic–pituitary–adrenal axis. The exact site of the defect has to be identified by further tests, e.g. the synacthen stimulation test.

See also: **human growth hormone**

INSULIN TEST MEAL

See: **Hollander test**

INTERNATIONAL UNIT (IU)

A unit of enzyme activity. It is the amount of enzyme which catalyses the reaction of one micromole of substrate per minute.

INTERSTITIAL FLUID

The fluid found between the tissue cells. It is separated from blood plasma by the capillary wall which acts as a semipermeable membrane allowing the passage of water and small molecules but not the larger molecules such as proteins. Together with blood plasma, it constitutes the extracellular fluid compartment.

INTERSTITIAL CELL-STIMULATING HORMONE (ICSH)

Synonym for luteinizing hormone.

See: **luteinizing hormone**

INTESTINAL DISACCHARIDASES

See: **disaccharidases and disaccharidase deficiency**

INTRACELLULAR FLUID

One of the two main fluid compartments into which the body can be theoretically divided, the other being the extracellular compartment.

INTRINSIC FACTOR

A mucoprotein, secreted by the parietal cells of the stomach, which binds with vitamin B_{12} (extrinsic factor) to form a complex. This complex binds to the intestinal cell and enables the vitamin to be absorbed. Lack of intrinsic factor can occur in various conditions leading to vitamin B_{12} deficiency and pernicious anaemia. Intrinsic factor secretion can be investigated by the Schilling test.

See also: **Schilling test, vitamin B_{12}**

INTRINSIC FACTOR ANTIBODIES

Autoantibodies which interfere with the actions of intrinsic factor and prevent the absorption of vitamin B_{12}. One type of antibody blocks the attachment of vitamin B_{12} to intrinsic factor (blocking antibody). Another type of antibody (binding antibody) binds to intrinsic factor regardless of whether vitamin B_{12} is present or not.

INULIN CLEARANCE TEST

Inulin is a plant polysaccharide which, when infused intravenously, is filtered at the glomerulus and excreted in the urine without reabsorption or secretion by the renal tubules. Estimation of its clearance is therefore a measure of the glomerular filtration rate. Inulin can be measured by the red colour it gives with resorcinol, following hydrolysis.

Further reading: Mitchell, F.L.. Veall, N. and Watts, R.W.E. (1972). Scientific Review No. 2. Renal function tests suitable for clinical practice. *Ann. Clin. Biochem.*, **9**, 1

IODINE

An essential element necessary for the synthesis of thyroid hormones. A dietary deficiency of iodine results in a compensatory thyroid enlargement in order to avoid hypothyroidism (a euthyroid goitre).

See also: **thyroxine**

ION-EXCHANGE CHROMATOGRAPHY

A chromatographic technique for the separation of a sample mixture based on differences of the ionic charge of each component. This determines their behaviour on an ion-exchange resin. These resins are large polymers such as cellulose or polystyrene having ionic groups incorporated into their structure.

Cation-exchange resins have negatively charged functional groups such as carboxylic acid (–COOH) or sulphonic acid ($-SO_3H$) groups. These are associated with loosely bound cations which can exchange with cations of the sample mixture.

Anion-exchange resins have positively charged functional groups such as quaternary ammonium groups. These exchange their loosely bound anions with anions of the sample mixture.

Ion-exchange chromatography has a wide variety of applications including the separation of proteins, hormones, enzymes and amino acids.

Further reading: General list of analytical textbooks

ION-SELECTIVE ELECTRODES

Certain membranes are permeable to selected ions. If such a membrane separates two solutions of a diffusible ion, and, if the concentration of the ion in one solution is greater than in the other, the membrane develops an electrical potential. If the solution on one side of the membrane is made the reference solution, and the solution on the other side is made the test solution, the membrane potential is directly proportional to the logarithm of the activity (for practical purposes, the concentration) of the diffusible ion in the test solution. Measurement of this potential therefore gives an estimate of the amount of that diffusible ion.

Three types of electrode are available:

(1) Glass electrodes. By varying the composition of the glass, the membrane can be made selective for H^+ (i.e. a pH electrode), K^+, Na^+, NH_4^+ and other ions.

(2) Solid state electrodes where the membrane can consist of an active substance which gives the membrane its specificity, e.g. AgCl for chloride measurement (used in sweat analysis) and AgBr for bromide measurement.

(3) Liquid ion-exchange electrodes where the membrane consists of a solvent in which is dissolved an ion-selective carrier, e.g. valinomycin which binds K^+ or dioctyl phosphate which binds calcium. This latter membrane is used for the measurement of ionized calcium in serum.

IONTOPHORESIS

This is the migration of small ions in an electric field. The technique is used in the collection of sweat for the diagnosis of cystic fibrosis. The drug, pilocarpine, is made to migrate into the skin by means of electrodes placed on the skin. This drug stimulates sweat production which can be collected on filter paper and analysed for sodium or chloride.

IRON

Distribution

About 70% of the total body iron is present in haemoglobin. The bulk of the remainder is present in the iron storage compounds, ferritin and haemosiderin, and smaller amounts are present in myoglobin, cytochromes and iron-containing enzymes. About 0.1% of the body's iron is present in the plasma and is carried by the iron transport protein, transferrin.

Absorption and excretion

Iron is absorbed mainly in the ferrous form in the upper small intestine, the rate of absorption being influenced by oxygen tension, rate of erythropoiesis and the size of the iron stores. Phytate and phosphate in the gut reduce absorption while gastric acid and reducing substances increase absorption.

Desquamation from the gut, and, in women, menstruation account for the bulk of the iron lost from the body.

Plasma iron

In both men and women, diurnal variations can occur, values being higher in the morning than in the evening. Increases in plasma iron levels are found when there is increased erythrocyte destruction as in haemolytic anaemias or when there is excessive iron intake, e.g. repeated blood transfusions, iron poisoning and excessive absorption of iron from the gut (haemochromatosis).

Decreased plasma iron levels are found in iron deficiency anaemia, pregnancy, renal failure and in malignancies and infections.

Plasma iron levels by themselves, however, give little information about the iron status of the body. It is usual to accompany plasma iron levels by determination of the serum iron-binding capacity (q.v.). Plasma ferritin estimation may also give an indication of the iron status.

Measurement of plasma iron

There are two approaches to plasma iron determination:

(1) Atomic absorption spectrophotometry (especially the flameless methods).

(2) Colorimetric methods. These involve the reaction of iron with a chelating agent to produce a coloured complex. Among these chelating agents are thiocyanate, dipyridyl, tripyridyl, *o*-phenanthroline, bathophenanthroline and ferrozine. There are two stages in these types of determination. The first stage consists of the disruption of the iron–protein complex and this is achieved using a strongly acid pH e.g. by hydrochloric acid. The free iron is isolated from the protein and then reacts with the colour reagent. With the exception of thiocyanate, all the colour reagents mentioned above react only with iron in the ferrous state and therefore reducing substances such as ascorbic acid must be present to maintain the iron as Fe^{2+}.

Copper is the main interfering agent in the colorimetric methods. Haemoglobin iron may also interfere.

Detection of iron in gastric aspirates

Iron tablets are particularly poisonous in children. Iron can be detected in gastric aspirates by its reaction with ferricyanide to give a blue colour.

Further reading: General list of analytical and clinical textbooks

IRON BINDING CAPACITY

Determination of plasma iron levels by themselves give little information about the state of the body's iron stores. Measurement of the iron binding capacity of the plasma is considered to be a more reliable indicator of the iron status of the body since it takes into account the degree of iron saturation of transferrin in the plasma. There are three ways of measuring the degree of saturation of the transferrin:

(1) Measurement of the transferrin by an immunochemical technique such as radial diffusion. If the plasma iron is known, the degree of saturation can be calculated.

(2) Total iron binding capacity of the plasma. The total iron binding capacity of the plasma (TIBC) equals plasma iron plus the unsaturated iron binding capacity (UIBC). An excess of inorganic iron is mixed with the plasma and the excess is removed using a resin or magnesium carbonate. Estimation of the remaining iron in the plasma gives the total iron binding capacity. Comparison of this with the original serum iron level gives an indication of the degree of iron saturation.

(3) Unsaturated iron binding capactiy of the plasma (UIBC). In this approach a known amount of excess inorganic iron is added to the plasma and the free (unbound) iron is determined in the supernatant. Subtraction of this from the amount of iron originally added gives the UIBC.

Further reading: General list of analytical and clinical textbooks

ISOCITRATE DEHYDROGENASE

A widely distributed cytoplasmic enzyme which catalyses the oxidative decarboxylation of isocitrate to α-oxoglutarate, a reaction which is accompanied by the reduction of NADP to NADPH. (A mitochondrial enzyme also exists which uses NAD as a cofactor.) Elevated serum levels are found in liver disease. It can be measured by following the reduction of NADP by ultraviolet spectrophotometry.

Further reading: Wilkinson, J.H. (1976). *The Principles and Practice of Diagnostic Enzymology*. (London: Edward Arnold)

ISOELECTRIC FOCUSING

A technique for the separation of protein mixtures by their electrophoretic migration through a medium in which a con tinuous pH gradient has been established. Each protein migrates to the point where the pH is equal to its isoelectric point when it ceases to move, i.e. becomes focused.

The pH gradient is established by means of ampholytes (molecules which carry negative and positive charges) incorporated into a support medium such as polyacrylamide gel. One end of the support medium is immersed in an acid solution, the other in a basic solution. Electrodes are placed at each end and the passage of the current causes the ampholytes to move into positions that result in a gradual pH gradient from one end of the support medium to the other.

Further reading: Latner, A.L. (1975). Isoelectric focusing in liquid and gels as applied to clinical chemistry. In Bodansky, O. and Latner, A.L. (eds.) *Advances in Clinical Chemistry*. Vol. 17, p. 193. (New York: Academic Press)

ISOENZYMES

Different forms of an enzyme which catalyse the same chemical reaction. Isoenzymes can originate from different organs, e.g. liver and bone alkaline phosphatase, brain and heart muscle creatine kinase. There are several approaches to the determination of the proportions of an isoenzyme mixture:

(1) Physical separation of the different isoenzymes by techniques such as electrophoresis or chromatography. Alkaline phosphatase fractionation is sometimes performed in this way.

(2) Different behaviour of the isoenzymes towards different substrates (e.g. the specificity of heart LDH for hydroxybutyrate), different inhibitors (e.g. formaldehyde stable acid phosphatase) or heat inactivation (e.g. placental alkaline phosphatase).

(3) By raising specific antiserum against each isoenzyme, enabling their immunochemical estimation. The major problem with this type of technique is cross-reactivity of the antiserum with other isoenzymes.

ISOLEUCINE

An amino acid found in increased levels in the blood, along with valine and leucine, in the inborn error of metabolism, maple syrup urine disease. This is due to a deficiency of branched chain oxoacid decarboxylase, an enzyme involved in the catabolism of the three branched chain amino acids, valine, leucine and isoleucine.

See also: **maple syrup urine disease**

ISOMALTASE

An intestinal disaccharidase which hydrolyses isomaltose (two glucose molecules linked by 1:6 linkages). Congenital or acquired isomaltase deficiency can occur, usually in association with sucrase deficiency.

See also: **disaccharidases and disaccharidase deficiency**

ISOTACHOPHORESIS

A type of zone electrophoresis in which ions migrate through a medium at equal speeds under the influence of an applied electrical current. An isotatchophoretic system can be established when two electrolytes are in contact with each other and sharing a common cation. The leading ion has a higher mobility and the trailing ion a lower mobility than any of the components in the sample. The mixture to be resolved is placed at the interface of the two ions where a voltage discontinuity exists. The application of an electrical current to the system results in a slower ion in the mixture being accelerated to the speed of the preceding ion. At equilibrium, the ion species form discrete bands which migrate through the gel at equal speeds. These zones reflect the absolute mobilities of the ions. The technique appears to have few applications in clinical chemistry.

J

JAFFÉ REACTION

The reaction of creatinine with picric acid to produce a red colour. It is used for the measurement of creatinine in biological fluids.

See also: **creatinine**

JAFFÉ TEST

A test for the detection of excess indican in urine based on its oxidation by hypochlorite to indigo blue which is extracted into chloroform.

See also: **indican**

JAUNDICE

A yellow coloration of the skin due to the presence of abnormal levels of bilirubin in the blood. Jaundice can occur as a result of:

(1) Increased amounts of bilirubin arriving at the liver, for instance in haemolytic conditions.

(2) A failure to conjugate bilirubin by the liver, e.g. physiological neonatal jaundice.

(3) A failure to excrete the conjugated bilirubin, e.g. obstruction of the common bile duct by a gall stone.

The two latter causes of jaundice are often found simultaneously in some liver disorders, e.g. hepatitis and cirrhosis.

Further reading: General list of clinical textbooks

JENDRASSIK AND GROF METHOD

A commonly used method for the estimation of total and conjugated bilirubin in serum. It is based on the diazotization of bilirubin by diazotized sulphanilic acid to azobilirubin which is coloured blue in alkaline solutions. Caffeine–sodium benzoate is used as the accelerating agent, and also included in the reaction mixtures is sodium acetate which buffers the pH of the diazotization reaction. Ascorbic acid, which destroys the excess diazo reagent, is used to terminate the reaction.

See also: **bilirubin**

K

KARMEN UNIT

A unit by which aminotransferase activity may be expressed. It is defined as the amount of enzyme in 1 ml of serum which will cause a change of 0.001 in the absorbance of NADH at 340 nm (1 cm light path) using appropriate reaction conditions.

KATAL

An enzyme unit by which it is recommended all enzymic activity is expressed. It is the amount of enzyme which catalyses the transformation of one mole of substrate per second under defined conditions.

KERATAN SULPHATE

A mucopolysaccharide excreted in the urine in some forms of mucopolysaccharidosis.

See: **mucopolysaccharidoses**

KETO-DIASTIX

A reagent stick manufactured by Ames, being a combination of Diastix (q.v.) and Ketostix (q.v.), for the simultaneous detection of glucose and ketones in urine.

Further reading: Kutter, D. (1977). *Rapid Clinical Diagnostic Tests.* (Munich–Vienna–Baltimore: Urban and Schwarzenberg)

17-KETOGENIC STEROIDS

See: **17-oxogenic steroids**

KETONE BODIES

A group of compounds which include acetoacetic acid, acetone and β-hydroxybutyrate. They result from fatty acid oxidation by the liver. Increased blood and urine levels are found in conditions such as diabetes mellitus or starvation, when fat oxidation is increased. Since hydrogen ions are produced with ketone bodies, a metabolic acidosis can result (diabetic ketoacidosis).

Detection of ketone bodies in urine

(1) *Rothera's test.* Ketone bodies react with sodium nitroprusside to produce a purple colour. This is the basis of the 'Acetest' tablet test and 'Ketostix' reagent strip test.

(2) *Gerhardt's test.* This is the reaction of ketone bodies with ferric chloride to produce a brownish-red colour.

17-KETOSTEROIDS

See: **17-oxosteroids**

KETOSTIX

A reagent stick manufactured by Ames for the detection o: ketones in serum and urine. It is based on the reaction of keton(bodies with nitroprusside to form a purple colour.

Further reading: Kutter, D. (1977). *Rapid Clinical Diagnosti(Tests.* (Munich–Vienna–Baltimore: Urban and Schwarzen burg)

KIDNEYS

The functions of the kidneys can be considered as:

(1) The elimination of the waste products of metabolism, e.ƒ urea.

(2) Homeostatic regulation of many substances by the ren: tubules, including water, electrolytes, bicarbonat(hydrogen ions, glucose and amino acids.

(3) It can be considered as an endocrine organ since produces renin, erythropoietin and converts 25-hydrox' cholecalciferol to the active 1,25-dihydroxychol(calciferol.

Further reading: General list of clinical textbooks

KING-ARMSTRONG UNIT

A unit by which acid and alkaline phosphatase activity can be expressed. A unit of alkaline phosphatase activity is the amount of enzyme which liberates 1 mg of phenol from phenol phosphate in 15 minutes at pH 10.0 and 37 °C. The acid phosphatase unit is similar except that the pH is 4.9 and the reaction time 60 minutes.

See also: **alkaline phosphatase**

KJELDAHL TECHNIQUE

A reference method for total protein determination. It consists of protein digestion with sulphuric acid, by which the nitrogen in the protein is converted to ammonium sulphate. Excess alkali is added and the liberated ammonia is distilled into excess standard acid and estimated by back titration with standard alkali. The protein content can then be calculated on the assumption that the proteins all contain 16% nitrogen by weight.

KOBER REACTION

A reaction in which oestrogens (Kober chromogens), when heated with sulphuric acid, yield an orange-yellow complex which can be estimated fluorimetrically or colorimetrically.

See also: **oestrogens**

KRABBE'S DISEASE (GLOBOID CELL LEUKODYSTROPHY)

An inborn error of lipid storage in which there is a deficiency of the enzyme galactocerebroside β-galactosidase. The disease becomes apparent in early infancy with blindness and deafness usually leading to death within two years.

Further reading: Suzuki, K. and Suzuki, Y. (1978). Galactosylceramide lipidosis. Globoid cell leukodystrophy (Krabbe's disease). In Stanbury, J.B., Wyngaarden, J.B. and Fredrickson, D.S. (eds.) *The Metabolic Basis of Inherited Disease*. 4th Edn., p. 747. (New York: McGraw-Hill)

L

LABSTIX

A dipstick test manufactured by Ames for the simultaneous detection of ketones, glucose, protein and haemoglobin in urine and also for the measurement of urinary pH.

LACTASE

An intestinal disaccharidase which cleaves lactose to yield glucose and galactose. Lactase deficiency can occur which may be congenital or acquired.

See also: **disaccharidases and disaccharidase deficiency**

LACTATE

Pyruvate, the end product of glycolysis, can be metabolized further in the Krebs cycle, or, if there is inadequate oxygen available, it can be converted to lactate.

Increased blood levels

Increased blood lactate can cause a metabolic acidosis (lactic acidosis). It occurs in conditions where there is poor tissue perfusion resulting in anoxia, e.g. circulatory failure and shock. Lactic acidosis can also occur in Type I glycogen storage disease and in phenformin treatment.

Measurement

Blood lactate can be measured using lactate dehydrogenase which converts the lactate to pyruvate, causing a reduction of NAD to NADH which can be measured spectrophotometrically.

Further reading: Krebs, H.A., Woods, H.F. and Alberti, K.G.M.M. (1975). Hyperlactataemia and lactic acidosis. In Marks, V. and Hales, C.N. (eds.) *Essays in Medical Biochemistry*. Vol. 1, p. 81. (London: The Biochemical Society and the Association of Clinical Biochemists)

LACTATE DEHYDROGENASE (LDH)

An enzyme which catalyses the reversible transformation of lactate to pyruvate:

$$\text{lactate} + \text{NAD} \rightleftharpoons \text{pyruvate} + \text{NADH}$$

It is found in many organs of the body including liver, heart, skeletal muscle and red cells. There are five isoenzymes of lactate dehydrogenase and tissues vary in the proportions of each isoenzyme they contain. The electrophoretically faster isoenzymes (LD_1 and LD_2) are found in heart muscle, while LD_5 is found predominantly in the liver.

Increased serum levels

Increases are found after myocardial infarction (reaching a peak value 48–72 hours after the event), haematological disorders, and liver and skeletal muscle disease.

Measurement

(1) Ultraviolet spectrophotometry. The enzyme can be measured in either direction, using either lactate as the substrate and measuring the increase in NADH, or pyruvate as the substrate in which case the decrease in absorbance is measured.

(2) There are several colorimetric methods available. The NADH formed in the forward reaction can be used to reduce a tetrazolium dye. Another method is based on the reaction of pyruvate, formed in the forward reaction, with 2,4-dinitrophenylhydrazine to form the coloured phenylhydrazone derivative.

(3) The isoenzymes associated with heart muscle can convert α-oxobutyrate (which has a similar structure to pyruvate) to α-hydroxybutyrate more effectively than the other LDH isoenzymes. Measurement of serum hydroxybutyrate dehydrogenase activity therefore gives an estimate of heart LDH (see separate entry on **hydroxybutyrate dehydrogenase**).

Further reading: Wilkinson, J.H. (1976). *The Principles and Practice of Diagnostic Enzymology*. (London: Edward Arnold)

LACTOGENIC HORMONE

See: **prolactin**

LACTOSE

A disaccharide composed of glucose and galactose. It is found in milk and milk products.

LACTOSE TOLERANCE TEST

A test used in the investigation of intestinal lactase deficiency. It consists of the oral administration of lactose followed by the collection of blood samples in which glucose is measured. If intestinal lactase is present, the lactose is broken down to glucose and galactose which are then absorbed. The blood glucose level should therefore increase. However, this does not occur if there is a deficiency of lactase. A 'flat' lactose tolerance test should, however, be followed by a glucose tolerance test in order to check that a generalized malabsorption does not exist.

See also: **disaccharidases and disaccharidase deficiency**

LAEVULOSE

See: **fructose**

LAG STORAGE CURVE

See: **glucose tolerance tests**

LAMBERT'S LAW

The amount of light absorbed by a solution is proportional to the length of the light path.

See also: **absorbance**

LANGE COLLOIDAL GOLD TEST

See: **colloidal gold reaction**

LASER NEPHELOMETRY

See: **nephelometry**

LAURELL 'ROCKETS'

See: **crossed electrophoresis**

LEAD AND LEAD POISONING

Lead poisoning can result from the use of lead drinking utensils and water pipes, or in children, it can occur as a result of eating lead paint. The symptoms of lead poisoning include intestinal cholic, neuropathy, encephalopathy and anaemia. It can be treated with chelating agents such as penicillamine and EDTA.

Diagnosis of lead poisoning

(1) Blood and urine lead levels.

Lead interferes with the biosynthesis of haem and this results in a build up of its precursors, e.g. δ-aminolaevulinic acid and coproporphyrin III. Thus other indirect tests of lead poisoning are:

(2) Urine δ-aminolaevulinic acid.

(3) Urine coproporphyrin.

(4) Red cell fluorescence (due to porphyrins).

(5) Decreased erythrocyte δ-aminolaevulinic acid dehydratase activity.

Measurement

(1) Colorimetric methods based on the reaction of lead with diphenylthiocarbazone to form a red complex.

(2) Atomic absorption spectrophotometry. Because of the low levels of lead, flameless methods must be used.

Further reading: Christian, G.D. (1976). The biochemistry and analysis of lead. In Bodansky, O. and Latner, A.L. (eds.) *Advances in Clinical Chemistry*. Vol. 18, p. 289. (New York: Academic Press)

Elder, G.H. (1976). Acquired disorders of haem synthesis. In Marks, V. and Hales, C.N. (eds.) *Essays in Medical Biochemistry*. Vol. 2, p. 75. (London: The Biochemical Society and the Association of Clinical Biochemists)

LECITHIN

A glycerophosphatide containing the amino alcohol, choline, and phosphatidic acid (a compound formed from glycerol, phosphate and fatty acids). It is needed for the normal surface activity of the lung. Lecithin is found in amniotic fluid where it originates from the respiratory tract of the fetus. This can be used to predict respiratory distress since low amnotic fluid lecithin levels are associated with lung dysmaturity. Higher levels are found only in the last few weeks of gestation and delivery at this stage has a favourable prognosis.

Measurement of lecithin

(1) Many laboratories measure the ratio of lecithin to sphingomyelin in amniotic fluid (L/S ratio). Sphingomyelin is another lipid found in amniotic fluid but, unlike lecithin, its level remains fairly constant throughout gestation. Low L/S ratios therefore indicate the likelihood of respiratory distress if the infant is delivered at that stage.
The measurement of L/S ratios entails extracting the lipids into an organic solvent and then separating them by thin layer chromatography. The lipids are stained and the intensity of the lecithin and sphingomyelin spots can be measured by densitometry.

(2) Lecithin can also be measured as lecithin itself, as lecithin phosphate or as palmitate (a constituent of lecithin).

See also: **palmitic acid**

Further reading: Lind, T. (1975). The analysis of amniotic fluid. *Br. J. Hosp. Med.,* **14**, 631

Whitfield, C.R. and Sproule, W.B. (1974). Fetal lung maturation. *Br. J. Hosp. Med.,* **12**, 678

Editorial. (1979). Biochemistry of surfactant. *Lancet,* **1**, 762

LECITHIN-CHOLESTEROL ACYL TRANSFERASE (LCAT)

An enzyme which catalyses the formation of cholesterol esters by the transfer of fatty acids from lecithin to the cholesterol of plasma lipoproteins (especially high density lipoproteins). A rare inborn error occurs in which there is a deficiency of this enzyme. Patients with this condition have high plasma levels of unesterified cholesterol.

Further reading: Gjone, E., Norum, K.R. and Glomset, J.A. (1978). Familial lecithin: cholesterol acyl transferase deficiency. In Stanbury, J.B., Wyngaarden, J.B. and Fredrickson, D.S. (eds.) *The Metabolic Basis of Inherited Disease.* 4th Edn., p. 589. (New York: McGraw-Hill)

LECITHIN-SPHINGOMYELIN RATIO (L/S RATIO)

See: **lecithin**

LESCH-NYHAN SYNDROME (JUVENILE HYPERURICAEMIA)

An inborn error of metabolism occuring in young male children the symptoms of which include aggressive behaviour accompanied by self-mutilation. It is due to a deficiency of hypoxanthine-guanine phosphoribosyl transferase (HGPRT), an enzyme involved in the recycling of hypoxanthine and other purines. A deficiency of this enzyme results in severe hyperuricaemia.

Further reading: Kelley, W.N. and Wyngaarden, J.B. (1978). The Lesch–Nyhan syndrome. In Stanbury, J.B., Wyngaarden, J.B. and Fredrickson, D.S. (eds.) *The Metabolic Basis of Inherited Disease.* 4th Edn., p. 1011. (New York: McGraw-Hill)

LEUCINE

A branched-chain amino acid found in elevated amounts in the blood along with valine and isoleucine in the inborn error of metabolism, maple syrup urine disease. This is due to a deficiency of branched-chain oxoacid decarboxylase, an enzyme involved in the degradation of the branched-chain amino acids.

See also: **maple syrup urine disease**

LEUCINE AMINOPEPTIDASE (LAP)

An enzyme which is found in elevated levels in the serum in hepatobiliary conditions.

Further reading: Wilkinson, J.H. (1976). *The Principles and Practice of Diagnostic Enzymology.* (London: Edward Arnold)

LEUCINE SENSITIVITY

Leucine is known to stimulate insulin secretion by the pancreas. Some individuals are particularly sensitive to leucine and develop hypoglycaemia. The condition can be diagnosed by giving oral leucine and measuring blood glucose and insulin levels in samples collected every fifteen minutes for one hour. In individuals with leucine sensitivity there is a sharp drop in the blood glucose and a rise in the plasma insulin levels.

LH/FSH-RELEASING HORMONE TEST

LH/FSH-releasing hormone normally stimulates the release of gonadotrophins from the pituitary. The test consists of injecting LH/FSH-releasing hormone and collecting blood samples for LH and FSH estimation. An impaired response occurs in hypopituitarism while an exaggerated response occurs in primary gonadal failure.

LIEBERMANN-BURCHARD REACTION

The reaction of cholesterol with acetic anhydride and sulphuric acid to give a green colour. This is the basis of some methods of cholesterol estimation.

See also: **cholesterol**

LIGHT-CHAIN DISEASE

Synonym for Bence Jones myeloma.

See: **myeloma**

LIPASE

An enzyme which catalyses the hydrolysis of glycerides into glycerol and fatty acids. Several types of lipase may be present in normal plasma, including pancreatic lipase and lipoprotein lipase. Increased serum levels are found in acute pancreatitis in much the same way as serum amylase.

Measurement

(1) By measuring the decrease in turbidity of a fat emulsion.

(2) By titrating with standard alkali the free fatty acids formed.

(3) Several colorimetric procedures have also been used.

LIPIDOSES

A group of inherited diseases in which there are deficiencies of specific lysosomal hydrolases. The diseases are characterized by the deposition of complex lipids. They include Gaucher's disease, Fabry's disease, Krabbe's disease, Tay–Sachs disease, generalized gangliosidosis, Niemann–Pick disease and metachromatic leukodystrophy.

See the separate entries for these diseases.

LIPIDS

A collective term which, in the context of clinical chemistry requests, usually means cholesterol, triglycerides and possibly lipoprotein analyses.

LIPOPROTEIN LIPASE (CLEARING FACTOR)

A plasma enzyme released from capillary endothelial cells. It catalyses the hydrolysis of protein-bound triglycerides. Increased serum enzyme activity is found after heparin injection.

LIPOPROTEINS

Complexes of lipid and protein which transport lipids in the blood. There are two main systms of nomenclature, one based on their electrophoretic mobility, the other on their behaviour during ultracentrifugation. These class the lipoproteins into four fractions:

(1) *Chylomicrons.* These contain mostly triglyceride with only a small proportion of protein. They are formed from triglycerides absorbed from the diet. They are transported to the tissues (especially adipose tissue) where they are broken down by the enzyme lipoprotein lipase.

(2) *Pre-β-lipoproteins* (very low density lipoproteins). These also are composed mainly of triglyceride but have higher proportions of cholesterol, phospholipid and protein than chylomicrons. Pre-β-lipoproteins transport endogenously synthesized triglycerides in the blood. Removal of triglycerides from pre-β-lipoproteins results in the formation of β-lipoproteins.

(3) *β-Lipoproteins* (low density lipoproteins). Of all the lipoproteins, these contain the highest proportion of

cholesterol. They are the major carriers of cholesterol in the blood.

(4) *α-Lipoproteins* (high density lipoproteins). These contain the largest proportion of protein. They appear to be involved in preventing the accumulation of cholesterol in tissues and, in this way, may be a factor in the prevention of coronary heart disease (see separate entry on **high density lipoproteins**).

There are several types of hyperlipoproteinaemias (see separate entry). Congenital deficiencies of lipoproteins can also occur.

See also: **abetalipoproteinaemia, Tangier disease**

Further reading: Miller, N.E. (1979). Plasma lipoproteins, lipid transport and atherosclerosis. Recent developments. *J. Clin. Pathol.,* **32**, 639

LIPOPROTEIN-X (LP-X)

A low density lipoprotein of abnormal composition, which can be detected in the serum of patients with cholestatic liver disease.

LITHIUM

Lithium carbonate is used in the treatment of manic depressive illnesses. The exact mechanism of its action is not known but it may interfere with cyclic-AMP mediated processes. The level of the drug in the blood must be monitored periodically as toxic reactions (tremors, diarrhoea, vomitting and drowsiness) can result from overdosage. Recently it has been found that lithium inhibits thyroxine biosynthesis and this may lead to hypothyroidism.

Lithium can be measured by either atomic emission or atomic absorption flame photometry.

LIVER

An organ whose biochemical functions include the following:

(1) Protein synthesis and breakdown, the latter resulting in the formation of urea.

(2) Synthesis of cholesterol, endogenous triglyceride, lipoproteins and phospholipids. It also catabolizes fatty acids and excretes the breakdown products of cholesterol metabolism in the bile.

(3) Gluconeogenesis, glycogenesis and, in fasting states, glycogenolysis occur in the liver.

(4) It is a site where many substances are stored, e.g. iron, vitamins and carbohydrate (as glycogen).

(5) It detoxicates and excretes many substances including bilirubin, cholesterol, steroid hormones and drugs.

Further reading: Triger, D. R. (1979). Physiological functions of the liver. *Br. J. Hosp. Med.,* **22**, 424

LIVER DISEASES

Two basic disorders of the liver can be considered as:

(1) Where the disorder is primarily that of liver cell destruction, e.g. hepatitis or cirrhosis.

(2) Where the disorder is primarily that of cholestasis which may be intrahepatic (e.g. biliary cirrhosis, cholangitis) or extrahepatic (e.g. gall stones obstructing the common bile duct, or carcinoma of the head of the pancreas).

LIVER FUNCTION TESTS

Strictly speaking these are tests which measure the specific functions of the liver, e.g. its excretory functions (bilirubin, BSP etc.) or its synthetic function (as reflected by serum albumin levels). In practice the term is also used to include tests for liver cell damage (e.g. aminotransferases released from damaged cells), tests for cholestasis (e.g. alkaline phosphatase and 5′-nucleotidase) and possibly a number of other miscellaneous tests (e.g. α-fetoprotein for primary hepatoma).

LLOYD'S REAGENT

An aluminium silicate adsorbant which can be used to isolate creatinine from interfering substances prior to its determination by the Jaffé reaction.

See also: **creatinine**

LONG-ACTING THYROID STIMULATOR (LATS)

An immunoglobulin (IgG) which stimulates the activity of the thyroid gland resulting in thyrotoxicosis.

See also: **hyperthyroidism**

LOW DENSITY LIPOPROTEINS

Synonym for β-lipoproteins.

See: **lipoproteins**

LUMINESCENT ASSAYS

Chemiluminescence is the emission of light by chemical processes: the energy of certain chemical reactions results in the production of molecules in an excited state. In returning to a lower energy state, light is emitted. (cf. fluorescence when the source of exciting energy is incident light.)

Several luminescent systems have been described, one of the more commonly used being the firefly luciferase system. This uses ATP as one of its substrates and light is emitted during the course of the chemical reaction. The enzyme can therefore be used in reactions where ATP is formed or consumed, e.g. creatine kinase assay, or the measurement of glycerol by glycerol kinase. Luminescent immunoassays, analogous to radioimmunoassay or enzyme-immunoassay have also been described, using luminol or luminol derivatives as the label.

Luminescence can be measured by modification of such instruments as photometers, fluorimeters and scintillation counters. Alternatively several instruments for the specific measurement of luminescence are commercially available.

Further reading: Gorus, F. and Schram, E. (1979). Review. Applications of bio- and chemiluminescence in the clinical laboratory. *Clin. Chem.,* **25**, 512

Whitehead, T.P. *et al.* (1979). Review. Analytical luminescence. Its potential in the clinical laboratory. *Clin. Chem.,* **25**, 1531

LUNDH TEST

A test of exocrine pancreatic function which consists of feeding a mixture of milk powder, corn oil and glucose to the patient and then measuring the trypsin in duodenal secretions in order to see if any stimulation of their secretion has taken place.

Further reading: Gowenlock, A.H. (1977). Scientific Review. No. 4. Tests of exocrine pancreatic function. *Ann. Clin. Biochem.,* **14**, 61

LUTEINIZING HORMONE (LH, INTERSTITIAL CELL STIMULATING HORMONE, ICSH)

A glycoprotein hormone secreted by the pituitary in response to a specific hypothalamic releasing factor (LH/FSH-RF). In males it stimulates testosterone secretion by the testis while in females it is responsible for ovulation. LH also acts with FSH in stimulating oestrogen secretion. In females plasma LH levels exhibit a sharp peak in the middle of the menstrual cycle, when ovulation occurs. Although LH can be measured by bioassay, it is now usually estimated by radioimunoassay.

Plasma LH

(1) This can be measured in women with irregular menstrual cycles to check if ovulation is occuring. This involves measuring the plasma levels in samples collected over several days in order to see if the mid-cycle ovulatory peak is present.

(2) In both males and females, low plasma LH levels indicate hypothalamic or pituitary dysfunction.

(3) In males, a high plasma LH level indicates primary testicular failure.

See also: **follicle-stimulating hormone, menstrual cycle**

LUTEINIZING HORMONE/FOLLICLE STIMULATING HORMONE RELEASING FACTOR (LH/FSH-RF)

A hypothalamic hormone responsible for regulating FSH and LH secretion by the pituitary.

LUTEOTROPHIC HORMONE (LUTEOTROPHIN)

See: **prolactin**

LYSINE

An amino acid excreted in large amounts in the urine in the inborn error, cystinuria. This condition is due to a failure of the tubular reabsorption mechanism for the dibasic amino acids, cystine, ornithine, arginine and lysine. Two types of hyperlysinaemia have also been described. In these two rare inborn errors, lysine accumulates in the blood.

See also: **cystinuria**

Further reading: Ghadimi, H. (1978). The hyperlysinaemias. In Stanbury, J.B., Wyngaarden, J.B. and Fredrickson, D.S. (eds.) *The Metabolic Basis of Inherited Disease*. 4th Edn., p. 387. (New York, McGraw-Hill)

LYSINE-VASOPRESSIN TEST

Lysine-vasopressin is a synthetic polypeptide which acts in a similar manner to corticotrophin releasing factor. It can therefore be used to assess the ability of the pituitary to secrete ACTH. In this test, lysine-vasopressin is injected intramuscularly and blood samples are collected for cortisol estimation. Failure of the cortisol level to rise significantly may indicate pituitary dysfunction.

LYSOZYME

A small molecular weight protein produced in large amounts in monocytic leukaemia. It passes into the urine where it can form a discrete band on protein electrophoresis. It is therefore sometimes mistaken for Bence-Jones protein. It has post-γ mobility and is well separated from the γ-globulin fraction.

M

MACROAMYLASAEMIA

A rare condition in which the serum amylase activity is raised due to the amylase combining with other plasma proteins to form complexes that cannot be filtered at the glomerulus.

See also: **amylase**

α_2-MACROGLOBULIN

A high molecular weight (820 000) plasma protein which migrates electrophoretically in the α_2-globulin region. It is retained in the blood in nephrotic syndrome, when the smaller molecular weight plasma proteins are lost in the urine, and, in this condition, levels may even be increased because of feedback stimulation of protein synthesis.

MACROGLOBULINAEMIA

A condition in which there are excess macroglobulins in the blood. The high levels of large molecular weight proteins lead to sluggish blood flow and this, in turn, can result in thrombosis in the small blood vessels. Retinal vein thrombosis, cerebral thrombosis and peripheral gangrene (hyperviscosity syndrome) are features.

Primary macroglobulinaemia

The most common cause is an IgM paraprotein (Waldenström's macroglobulinaemia). Macroglobulinaemia can also result from polymerization of IgG or Bence Jones protein to form large complexes.

Secondary macroglobulinaemia

This can be found in conditions where there is a polyclonal immunoglobulin response.

MACROGLOBULINS

A general term for the high molecular weight plasma proteins, e.g. α_2-macroglobulin and IgM.

MAGNESIUM

The metabolism and actions of magnesium are closely linked to those of calcium. It is present in bone where it enters and leaves along with calcium. It is also one of the major intracellular cations and tends to enter or leave cells under the same conditions that affect potassium movement.

Low serum magnesium levels

This causes symptoms similar to those of hypocalcaemia, e.g. tetany. The causes of hypomagnesaemia include:

(1) Diarrhoea, when excess magnesium is lost.

(2) Some hypocalcaemic conditions, e.g. hypoparathyroidism, when it moves out of bone along with calcium.

(3) Certain conditions which cause hypokalaemia such as primary aldosteronism.

High serum magnesium levels

Like high serum calcium levels, hypermagnesaemia can cause muscular hypotonia. Renal failure is the commonest cause of hypermagnesaemia.

Measurement

(1) Atomic absorption spectrophotometry is commonly used.

(2) It can be estimated fluorimetrically by its reaction with 8-hydroxy-5-quinoline sulphonic acid to form a fluorescent chelation complex.

(3) It can be determined colorimetrically by its reaction with dyes, e.g. titan yellow, Mann–Yoe dye, to form coloured complexes.

Further reading: Vernon, W.B. and Wacker, W.E.C. (1978). Magnesium metabolism. In Alberti, K.G.M.M. (ed.) *Recent Advances in Clinical Biochemistry*. Vol. 1, p. 39. (Edinburgh, London and New York: Churchill-Livingstone)

MALATE DEHYDROGENASE

An enzyme which catalyses the reaction:

$$\text{malate} + \text{NAD} \rightleftharpoons \text{oxaloacetate} + \text{NADH}$$

High levels are found in heart and skeletal muscle, liver, kidney and erythrocytes. Elevated serum concentrations are found in haemolytic conditions, liver disease and following myocardial infarction.

Further reading: Wilkinson, J.H. (1976). *The Principles and Practice of Diagnostic Enzymology.* (London: Edward Arnold)

MALIGNANT HYPERPYREXIA

A hereditary condition, characterized by muscular rigidity, acidosis and high body temperatures occuring after a general anaesthetic. Some susceptible individuals have raised plasma CPK levels.

Further reading: Wilkinson, J.H. (1976). *The Principles and Practice of Diagnostic Enzymology.* (London: Edward Arnold)

MALTASE

An intestinal disaccharidase which hydrolyses maltose into two glucose molecules. An acquired deficiency of the enzyme, along with other disaccharidases, can occur when there is generalized disease of the intestinal wall.

See also: **disaccharidases and disaccharidase deficiency**

MANNITOL

The determination of mannitol clearance can be used as an indication of the glomerular filtration rate, since it is filtered at the glomeruli and neither secreted nor reabsorbed by the renal tubules.

MAPLE SYRUP URINE DISEASE (BRANCHED-CHAIN KETONURIA)

An inborn error of metabolism in which there is a deficiency of the enzyme which decarboxylates the oxo acids that result from the breakdown of the three branched-chain amino acids, leucine, valine and isoleucine. The branched-chain oxo acids and amino acids pass out into the urine and give it a characteristic odour. Mental retardation and neurological symptoms occur and death at an early age results.

The disease can be diagnosed biochemically by the reaction of the urinary branched-chain oxo acids with dinitrophenylhydrazine to form characteristic dinitrophenylhydrazone derivatives which can be identified chromatographically.

Further reading: Dancis, J. and Levitz, M. (1978). Abnormalities of branched-chain amino acid metabolism (hypervalinaemia, maple syrup urine disease, isovaleric acidaemia and β-methylcrotonic aciduria). In Stanbury, J.B., Wyngaarden, J.B. and Fredrickson, D.S. (eds.) *The Metabolic Basis of Inherited Disease*. 4th Ed., p. 397. (New York: McGraw-Hill)

MAROTEAUX-LAMY SYNDROME

A form of mucopolysaccharidosis.

See: **mucopolysaccharidoses**

MASS SPECTROMETRY

A technique used primarily for the identification of molecules The sample is vapourized and the resulting vapour molecules ar broken down into charged fragments, either by bombardmen with a beam of electrons or other means. A mass analyse separates the charged particles by, for example, their deflectio along a circular path in a magnetic field. The ions pass throug narrow slits where they are detected and recorded on a char A compound has its own particular fragmentation patter enabling the identification of unknown molecules.

Mass spectrometry can be used in conjunction with ga chromatography which initially separates a mixture of unknow compounds.

Further reading: Roboz,J. (1975). Mass spectrometry in clinic chemistry. In Bodansky, O. and Latner, A.L. (eds.) *Ad vances in Clinical Chemistry*. Vol. 17, p. 109. (New Yor Academic Press)

M-BAND

A term which is used to describe a serum paraprotein of any immunoglobulin class (not just IgM). The M can stand for malignancy, myeloma or macroglobulinaemia.

See also: **paraprotein**

McARDLE'S DISEASE

A glycogen storage disease in which there is an absence of muscle phosphorylase activity.

See also: **glycogen storage diseases**

MEDIAN

This is the value which, in a distribution curve, divides the number of observations into two equal parts. In a normal distribution curve it is the same as the mode and the mean. In a skewed distribution it is different.

MEGALOBLASTIC ANAEMIA

A form of anaemia which can be due to vitamin B_{12} or folate deficiencies, these vitamins being required for nucleic acid synthesis.

See also: **folic acid, vitamin B_{12}**

MELANIN

A pigment produced by melanocytes which causes darkening of the skin. In patients with melanoma, a colourless precursor (melanogen) is excreted in urine, which darkens on standing due to the oxidation of melanogen to melanin.

MELANOCYTE-STIMULATING HORMONE (MSH)

A hormone secreted by the pituitary which causes increased melanin synthesis and darkening of the skin. Two forms of this hormone are known and are referred to as α-MSH and β-MSH. Both have structural similarities to ACTH. MSH secretion parallels ACTH secretion, being greatest when the circulating plasma cortisol level is low, as in Addison's disease. MSH secretion may also be under the control of hypothalamic release and inhibiting factors.

MELANOGEN

A colourless precursor of the pigment melanin. In patients with melanoma, it is excreted in large amounts in the urine, where it is oxidized on standing to melanin producing a dark urine. The screening tests for melanogen should therefore be performed on fresh urine. Among the tests for melanogen detection are:

(1) Oxidation of the melanogen by ferric chloride to produce a black precipitate of melanin.

(2) Addition of sodium nitroprusside to produce a blue colour (Thormählen's test).

MELATONIN

A hormone derived from the pineal body which lightens the colour of melanocytes in the frog and which blocks the action of MSH. In humans it acts on the brain and influences processes such as sleep, ovulation and puberty. Melatonin-secreting pinea tumours are associated with reduced gonadal function. Alternatively pineal destruction can cause precocious puberty.

MENINGITIS

Estimation of CSF proteins and glucose can be of assistance in diagnosing this condition.

See also: **cerebrospinal fluid, colloidal gold reaction**

MENKES' KINKY HAIR DISEASE

This is a progressive central nervous system degenerative disorder, having an X-linked recessive mode of inheritance. Th scalp hair is sparse and stubbly and when examined under th microscope, the hairs appear twisted and have partial break: Affected individuals have disturbed copper homeostasis. Th basic biochemical abnormality may be defective intestinal ab sorption of copper.

Further reading: Sass-Kortsak, A. and Bearn, A. G. (1978 Hereditary diseases of copper metabolism. In Stanbury,

B., Wyngaarden, J. B. and Fredrickson, D. S. (eds) *The Metabolic Basis of Inherited Disease.* 4th Edn., p.1098. (New York: McGraw-Hill)

MENSTRUAL CYCLE

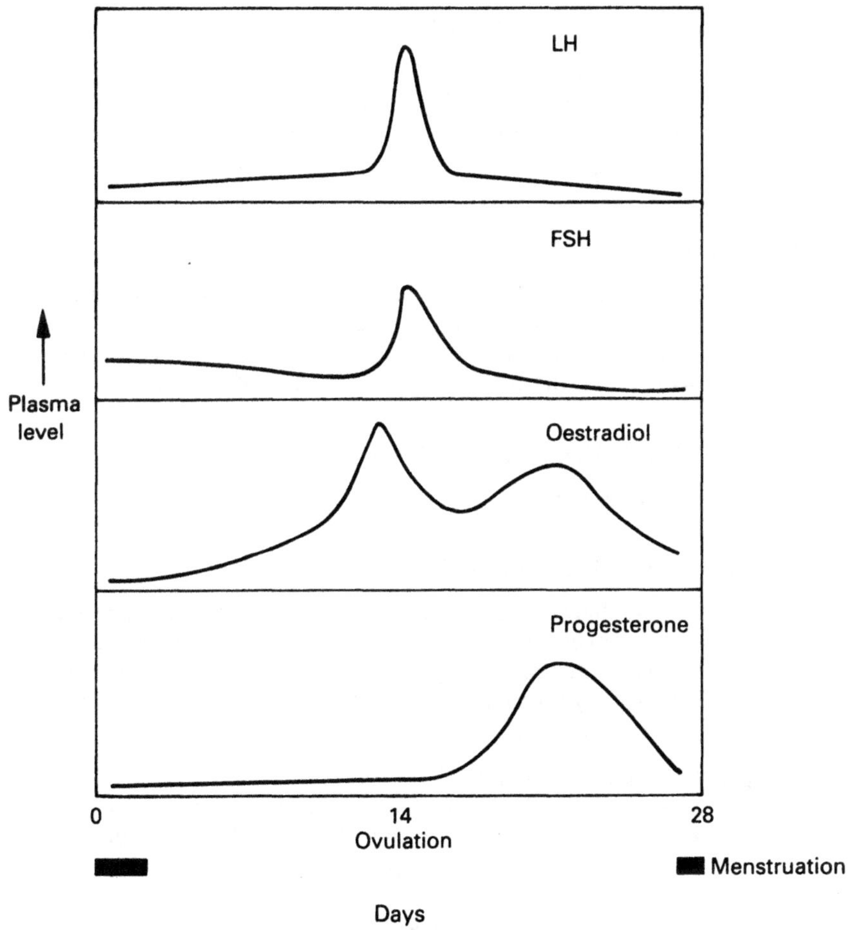

In a normal menstrual cycle the following hormonal changes occur:

(1) In the first half of the cycle (the follicular phase) FSH

stimulates the development of an ovarian follicle. Together with LH it stimulates the follicle to secrete oestradiol which reaches a peak level immediately prior to ovulation.

(2) At ovulation there is a burst of FSH and LH secretion which stimulates the follicle to release the ovum. The ruptured follicle is converted into a corpus luteum which secretes both progesterone and oestradiol during the second half of the cycle (the luteal phase).

(3) If fertilization does not occur, the corpus luteum involutes, the progesterone and oestradiol levels fall and the endometrium breaks down with menstruation resulting.

Further reading: General list of clinical textbooks

MERCURY

A heavy metal which, if present in the body in large amounts, can result in renal disease. It can be measured by atomic absorption spectrophotometry or colorimetrically by its reaction with dithizone to give a pink coloured dithizonate.

METABOLIC BALANCE STUDIES

See : **balance studies**

METACHROMATIC LEUKODYSTROPHY

A lipidosis, characterized by the accumulation of sulphuric acid esters of cerebrosides in nervous tissue, due to a deficiency of the degradative enzyme, cerebroside sulphatase. Progressive paralysis and hypotonia are among the clinical findings resulting in death after a few years.

Further reading: Moser, H.W. and Dulaney, J.T. (1978) Sulphatide lipidosis. In Stanbury, J.B., Wyngaarden, J.B. and Fredrickson, D.S. (eds.) *The Metabolic Basis of Inherited Disease.* 4th Edn., p. 770. (New York: McGraw-Hill)

METADRENALINE (METANEPHRINE)

A methylated metabolite of adrenaline. It can be measured, along with normetadrenaline, the methylated derivative of noradrenaline, in the investigation of catecholamine secreting tumours. They can be isolated by column chromatography and then converted to vanillin which is then estimated spectrophotometrically.

See also: **catecholamines**

METANEPHRINE

See: **metadrenaline**

METHAEMALBUMIN

A combination of haematin (oxidized haem, containing iron in the ferric state) and albumin. It is found in the blood when there is intravascular haemolysis and in cases of acute haemorrhagic pancreatitis. It can be identified in blood by its spectral characteristics or by Schumm's test, in which an albumin haemochromogen, which has a distinctive absorption spectrum, is formed.

METHAEMOGLOBIN

Haemoglobin which has been oxidized and contains iron in the ferric state. As a result, it cannot act as an oxygen carrier and this leads to cyanosis. Methaemoglobin is produced continually in red blood cells but is reconverted to haemoglobin by methaemoglobin reductases. Some oxidizing drugs, e.g. phenacetin, interfere with this regulation and methaemoglobinaemia can result. A congenital methaemoglobinaemia due to a deficiency in methaemoglobin reductase has also been found. Methaemoglobin can be identified in plasma by its spectral characteristics, which change when a reducing agent, such as dithionite, is added.

METHANOL

An alcohol occasionally encountered in cases of poisoning. It is metabolized to formaldehyde and formic acid resulting in a metabolic acidosis. A number of other toxic effects can result including blindness. It can be estimated by gas–liquid chromatography or colorimetrically by oxidation to formaldehyde followed by its reaction with chromotropic acid.

METHAQUALONE (MANDRAX)

A hypnotic drug which may lead to addiction. It can be estimated in body fluids by ultraviolet spectrophotometry following solvent extraction.

Further reading: Yeoman, W.B. (1971). Toxicological analysis in the clinical chemistry laboratory. *Ann. Clin. Biochem.*, **8**, 93.

Meade, B.W. *et al.* (1972). Technical Bulletin No. 24 Simple tests to detect poisons. *Ann. Clin. Biochem.*, **9**, 35

METHYLAMINE TEST (FEARON'S TEST)

A test for the detection of lactose in urine based on its reaction with methylamine to give a red colour.

METHYLENE BLUE

A dye which is found in certain medicines. It can result in the passage of a greenish-blue urine.

METHYLMALONIC ACID

A compound which can be found in the urine of patients with vitamin B_{12} deficiency and in the very rare inborn error of metabolism, methylmalonic aciduria. It is an intermediate in the metabolism of propionic acid (itself a metabolite of certain amino acids, particularly valine and isoleucine). Vitamin B_{12} is a cofactor in the enzymic step by which methylmalonyl coenzyme A is converted to succinyl coenzyme A. In vitamin B_{12} deficiency methylmalonate accumulates and passes out into the urine. Its measurement in urine can therefore be used to diagnose deficiency of this vitamin. It can be estimated colorimetrically by its reaction with diazotized *p*-nitroaniline to form a green compound.

See also: **methymalonic aciduria**

Further reading: Chanarin, I. (1977). Folates, cobalamins and their interrelationship in man. In Marks, V. and Hales, C.N. (eds.) *Essays in Medical Biochemsistry*. Vol. 3, p. (London: The Biochemical Society and the Association of Clinical Biochemists)

METHYLMALONIC ACIDURIA

A rare inborn error of metabolism in which methylmalonic acid is excreted in the urine. It can be due to a deficiency of the enzyme, methymalonate coenzyme A mutase, which converts methylmalonate CoA to succinyl CoA. Another variant of the disease is due to defective biosynthesis of the vitamin B_{12} coenzyme required for the mutase reaction.

Further reading: Rosenberg, L.E. (1978). Disorders of propionate, methylmalonate and cobalamin metabolism. In Stanbury, J.B.. Wyngaarden, J.B. and Fredrickson, D.S. (eds.) *The Metabolic Basis of Inherited Disease.* 4th Edn., p. 411. (New York: McGraw-Hill)

N-METHYLNICOTINAMIDE

A metabolite of nicotinic acid which can be measured in urine to diagnose nicotinamide deficiency. It can be measured by its reaction with ketones to give a fluorescent compound.

See also: **nicotinamide**

METOPIRONE TEST

See: **metyrapone test**

METYRAPONE (METOPIRONE) TEST

An *in vivo* test used to establish both the aetiology of Cushing's syndrome and in the investigation of adrenocortical hypofunction. Metyrapone inhibits 11-hydroxylase activity, one of the enzymes involved in the biosynthesis of cortisol. The reduced plasma cortisol levels should normally result in increased ACTH production which leads to increased synthesis of the cortisol precursor 11-deoxycortisol and its urinary metabolites. The drug is given orally and adrenal function can be followed by either measuring plasma 11-deoxycortisol levels or by measuring 17-oxogenic steroids in 24 hour urine samples. A diminished response could indicate failure of the pituitary to secrete ACTH or an adrenal carcinoma or adenoma. An exagerated response is found in adrenal hyperplasia.

Further reading: Hall, R., Anderson, J., Smart, G.A. and Besser, M. (1974). *Fundamentals of Clinical Endocrinology.* 2nd Edn. (London: Pitman Medical Publishing Co.)

MICELLES

Aggregates of bile salts, monoglycerides and fatty acids formed during the digestion of fat. They serve to solubilize the lipids in order that they can be absorbed.

β_2-MICROGLOBULIN

A small molecular weight plasma protein. Increased urinary excretion of β_2-microglobulin occurs in renal tubular proteinurias.

MILK-ALKALI SYNDROME

A rare form of hypercalcaemia that may result from large intakes of milk along with alkalis during the treatment of peptic ulcers.

MILLON REACTION

The reaction of tyrosine or tyrosine-containing proteins with mercuric nitrate to produce a red colour. It can be used to detect tyrosine in urine as in tyrosinosis.

MINERALOCORTICOIDS

Adrenocortical steroid hormones which influence sodium and potassium transport across cell membranes. This is particularly important in the renal tubules where they promote sodium reabsorption and potassium excretion. Aldosterone is the main adrenal mineralocorticoid. 11-Deoxycorticosterone and 11-deoxycortisol also have mineralocorticoid activity.

See also: **aldosterone**

MITOCHONDRIAL ANTIBODIES

Autoantibodies found in the serum of the majority of patients with primary biliary cirrhosis.

MODE

In a distribution curve, this is the value which occurs most commonly. In a normal distribution it is the same as the median and the mean. In a skewed distribution it is different.

MOLALITY

The number of moles of solute per kilogram of solvent.

MOLARITY

The number of moles of solute per litre of solution.

MOLECULAR SIEVE CHROMATOGRAPHY

See: **gel filtration**

MOLYBDENUM

This is a component of xanthine oxidase and other metallo-enzymes.

MONOCHROMATIC LIGHT

Light containing only a narrow part of the spectrum. It can be produced from a hollow cathode lamp as in atomic absorption spectrophotometry or by the use of diffraction gratings, prisms and filters to isolate a specific spectral region from a tungsten, hydrogen or other lamp.

MONOCLONAL GAMMOPATHY

A term used to describe diseases in which an abnormal protein, resulting from the proliferation of a single type of immuno-globulin-producing cell (i.e. a single clone), is found in the serum, usually in the γ-globulin region. Myeloma and macro-globulinaemia are the commonest monoclonal gammopathies.

See also: **macroglobulinaemia, myeloma**

MONO-IODOTYROSINE

An intermediate in the formation of thyroxine.

See: **thyroxine**

MORQUIO'S SYNDROME

A form of mucopolysaccharidosis.

See: **mucopolysaccharidoses**

MUCOPOLYSACCHARIDES

These are heteropolysaccharides composed of equal amounts of an amino sugar and a uronic acid, alternately linked through a glycosidic bond. Examples of mucopolysaccarides are chondroitin sulphate, keratan sulphate and dermatan sulphate. They are consituents of connective tissue. Mucopolysaccarides are excreted in the urine in large amounts in the inborn errors of metabolism known as the mucopolysaccharidoses.

Detection of urinary mucopolysaccharides

Several screening tests are available based on the precipitation of the negatively charged acid mucopolysaccharides by a large positively charged organic ion. These include:

(1) Measurement of the turbidity after addition of a bovine albumin solution.

(2) Measurement of the turbidity after addition of a cetylpyridinium choloride solution.

(3) The Alcian Blue test. Urine is spotted on to filter paper and immersed in a solution of Alcian Blue. If mucopolysaccharides are present, the urine spot stains blue.

If these screening tests are positive, the mucopolysaccharides can be identified by electrophoresis.

See also: **mucopolysaccharidoses**

Further reading: Kennedy, J.F. (1976). Chemical and biochemical aspects of the glycosaminoglycans and proteoglycans in health and disease. In Bodansky, O. and Latner, A.L. (eds.) *Advances in Clinical Chemistry*. Vol. 18, p. 1. (New York: Academic Press)

MUCOPOLYSACCHARIDOSES

A group of inherited diseases in which there is abnormal metabolism and excretion of the various mucopolysaccharides. The features of these diseases are a reflection of the inability of the body to link the mucopolysaccharides with proteins to form the basic substance of connective tissue. Skeletal abnormalities result from this. Affected individuals may also have grotesque facial expressions. The mucopolysaccharidoses have been classified as follows:

IH Hurler (Gargoylism)
IS Scheie
II Hunter
III Sanfilippo
IV Morquio
VI Maroteaux-Lamy

The distinction between the types depends on the particular lysosomal enzyme which is deficient. The commonest types are IV and IH.

The mucopolysaccharidoses can be diagnosed by the detection and identification of the excess mucopolysaccharides in the urine.

Further reading: McKusick, V.A., Neufeld, E.F. and Kelley, T.E. (1978). The mucopolysaccharide storage diseases. In Stanbury, J.B. Wyngaarden, J.B. and Fredrickson, D.S. (eds.) *The Metabolic Basis of Inherited Disease*. 4th Edn., p. 1282. (New York: McGraw-Hill)

MUCOPROTEINS (SEROMUCOIDS)

These are protein–carbohydrate compounds which contain 10–75% of carbohydrate and over 4% of hexosamines (as distinct from glycoproteins which contain from a trace up to 15% carbohydrate and less than 4% of hexosamines). Orosomucoid, haptoglobin and haemopexin are examples of mucoproteins. Increased serum mucoprotein levels are found in many inflammatory conditions such as acute infections and rheumatoid arthritis. Mucoproteins can be estimated by selective precipitation using phosphotungstic acid, followed by determination of the protein or carbohydrate content of the precipitate.

MUCOVISCIDOSIS

See: **cystic fibrosis**

MULTIPLE ENDOCRINE ADENOMATOSIS

A disorder in which there are adenomas or hyperplasia occuring simultaneously in the pituitary, adrenals, pancreas, parathyroids or in any combination of these glands. The symptoms therefore include hyperparathyroidism, acromegaly, Cushing's syndrome, gastric ulceration (if the pancreatic cells produce gastrin) and hypoglycaemia (if the pancreatic cells produce insulin).

Further reading: Hall, R., Anderson, J., Smart, G.A. and Besser, M. (1974). *Fundamentals of Clinical Endocrinology*. 2nd Edn., (London: Pitman Medical Publishing Co.)

MULTIPLE MYELOMA

See: **myeloma**

MULTIPLE SCLEROSIS

Estimation of CSF proteins can be of assistance in diagnosing this condition. One frequently used investigation is the measurement of IgG and albumin in CSF and plasma. Affected individuals have a higher than normal ratio of

$$\frac{\text{CSF IgG}}{\text{plasma IgG}} \quad \text{to} \quad \frac{\text{CSF albumin}}{\text{plasma albumin}}$$

indicating local IgG secretion.

See also: **cerebrospinal fluid, colloidal gold reaction**

Further reading: Hughes, R.A.C. (1977). Immunological aspects of multiple sclerosis *Br. J. Hosp. Med.*, **18,** 467

Editorial (1977). Laboratory tests for multiple sclerosis. *Lancet*, **2**, 131

MUREXIDE TEST

A test which is used for the detection of uric acid in calculi. It is based on the reaction of uric acid with nitric acid, followed by the addition of ammonium hydroxide to give ammonium purpurate which is coloured purple.

MUSCLE DISEASES

In muscle disease, such as the muscular dystrophies, enzymes leak from the damaged cells into the circulation where they can be measured. The enzymes most commonly measured in the diagnosis of muscle diseases are creatine kinase, aldolase, and aspartate and alanine aminotransferases.

Further reading: Pennington, R.J. (1971). Biochemical aspects of muscle disease. In Bodansky, O. and Latner, A.L. (eds) *Advances in Clinical Chemistry*. Vol. 14, p. 410. (New York: Academic Press)

MYELOMA (MULTIPLE MYELOMA, MYELOMATOSIS)

A condition, usually occuring in later life, in which there is a malignant proliferation of plasma cells of a single type in the bone marrow. The proliferation of the plasma cells results, in the great majority of cases, in the appearance in the serum of a paraprotein, which is of a single class and type of immunoglobulin. IgG and IgA myelomas are the most frequently occuring. IgM, IgD and IgE myelomas are rarer (the finding of an IgM paraprotein in the serum usually indicates macroglobulinaemia rather than an IgM myeloma). Bence Jones myelomas, in which only light chains are produced, are also known.

The symptoms of myeloma include bone pain, fractures, anaemia and infections. Cytotoxic drugs and radiotherapy are used for treatment.

Laboratory diagnosis

(1) The detection and identification of a paraprotein.

(2) The finding of Bence Jones protein in urine.

(3) The finding of malignant plasma cells in a bone marrow aspirate.

(4) Other biochemical findings include hypercalcaemia and the biochemical features of renal failure.

See also: **paraprotein**

MYOCARDIAL INFARCTION

Serum enzyme estimations can be of assistance in the diagnosis of myocardial infarction. Creatine kinase levels reach a peak value 24–30 hours after the infarction while, aspartate aminotransferase reaches a peak value around 36 hours after the event. Lactate dehydrogenase (or hydroxybutyrate dehydrogenase) reaches peak levels 2–3 days after the infarction.

Further reading: Wilkinson, J.H. (1976). *The Principles and Practice of Diagnostic Enzymology*. (London: Edward Arnold)

MYOGLOBIN

The oxygen binding protein of heart and skeletal muscle. Its molecular weight is one quarter that of haemoglobin and it resembles a haemoglobin subunit. It is released from muscle in crushing injuries and subsequently appears in the urine (myoglobinuria) where it imparts a dark colour. It can be identified in urine by its electrophoretic mobility or by its spectral characteristics. Radioimmunoassays have also been developed.

MYXOEDEMA

See: **hypothyroidism**

N

NATELSON MICROGASOMETER

An instrument for the measurement of total CO_2 in blood. Acid is added to the specimen and the CO_2 liberated is measured manometrically using the instrument.

See also: **bicarbonate**

NELSON'S SYNDROME

This is a condition which results from a pituitary adenoma, occuring after total adrenalectomy for Cushing's disease, due to bilateral adrenal hyperplasia. ACTH and MSH may be secreted in response to increased secretion of CRF by the hypothalamus and this results in skin pigmentation. The increased secretion of CRF may be an attempt to maintain the high levels of cortisol that existed before the operation.

NEPHELOMETRY

A technique for the measurement of light scattered by a suspension of particles. The amount of light scattered is proportional to the number and size of the particles. Thus the light is measured at an angle to the direction of the incident light (cf. turbidimetry, when transmitted light is measured). Nephelometric techniques are used for the measurement of lipoproteins and proteins (the latter by their reaction with a specific antibody to form immune complexes).

A recent development of this technique has been the use of a laser beam instead of a conventional light beam (laser nephelometry). Laser light has three advantages which make it suitable for use in light scattering techniques.

(1) It is very monochromatic.

(2) It has a high intensity.

(3) It has a high degree of collimation.

See also: **automated immune precipitation**

Further reading: White, P.A.E. and Strong, R. (1979). Automated immunoprecipitation and laser nephelometry. In Milford Ward, A. and Whicher, J.T. (eds.) *Immunochemistry in Clinical Laboratory Medicine.* p. 23. (Lancaster: MTP Press)

NEPHROGENIC DIABETES INSIPIDUS

A rare inborn error in which the renal tubules cannot respond to antidiuretic hormone. Large volumes of urine are passed and the patient becomes dehydrated.

Further reading: Andreoli, T.E. and Schafer, J.A. (1978) Nephrogenic diabetes insipidus. In Stanbury, J.B. Wyngaarden, J.B. and Fredrickson, D.S. (eds.) *The Metabolic Basis of Inherited Disease.* 4th Edn., p. 1634 (New York: McGraw-Hill)

NEPHROTIC SYNDROME

A condition in which there is extensive protein loss in the urine due to increased glomerular permeability. The glomerular disorder may be a primary condition (e.g. some types of glomerulonephritis) or secondary to other conditions, such as amyloidosis, diabetes mellitus and systemic lupus erythematosus. The resultant hypoalbuminaemia causes oedema.

Biochemical features of nephrotic syndrome

(1) Proteinuria. In mild cases only the lower molecular weight proteins such as transferrin and albumin are excreted in the urine (i.e. a high selectivity), but in more severe cases higher molecular weight proteins such as IgG are extensively excreted. Determination of the differential protein clearance gives an indication of the severity of the condition. This is the ratio of the clearance of a high molecular weight protein (e.g. IgG) to a lower molecular weight protein (e.g. transferrin). The lower the ratio, the greater the selectivity and the more favourable the prognosis, these cases responding better to treatment with steroids or cyclophosphamide.

(2) Hypercholesterolaemia with increases in the serum β-lipoprotein level.

Further reading: General list of clinical textbooks

NESSLERIZATION

The reaction of ammonia or ammonium compounds with an alkaline solution of mercuric and potassium iodides to give a yellow colour. This reaction can be used to estimate urea, when used in conjunction with the enzyme urease which converts urea to ammonia.

See also: **urea**

NEUROBLASTOMA

A catecholamine secreting tumour of sympathetic nervous tissue occurring in children. Approximately 40% occur in the adrenal medulla, the remainder being found in extra-adrenal sites.

NEUTRAL FAT

A general term applied to triglycerides.

See: **triglycerides**

NIACIN

A term which includes nicotinic acid and nicotinamide.

See: **nicotinamide**

NICOTINAMIDE

A member of the water-soluble B group of vitamins. It can be obtained from the diet or it can be synthesized endogenously from nicotinic acid, which is itself derived from tryptophan. Nicotinamide is a constituent of the coenzymes NAD and NADP which have widespread roles in intermediary metabolism. Deficiency of the vitamins causes pellagra. Patients with Hartnup's disease can develop a pellagra type condition probably due to insufficient endogenous synthesis of the vitamin from tryptophan.

Measurement

Nicotinic acid can be measured as an indication of nicotinamide deficiency. It can be estimated by its colorimetric reaction with cyanogen bromide and an aromatic amine. Nicotinic acid can also be measured as its urinary metabolite, *N*-methylnicotinamide, which reacts with ketones to give a fluorescent compound.

Further reading: General list of analytical and clinical textbooks.

NICOTINIC ACID

Intestinal bacteria can metabolize tryptophan to nicotinic acid. This can be converted by the body to nicotinamide.

See also: **nicotinamide**

NIEMANN-PICK DISEASE

A lipidosis characterized by the presence of increased amounts of sphingomyelin in the tissues. Four types (A–D) have been recognized. In types A and B, a deficiency of the lysozomal enzyme, sphingomyelinase, has been demonstrated. Type A affects the nervous system and carries a poor prognosis. Splenomegaly is the main feature of type B and affected individuals may survive to adulthood.

Further reading: Brady, R.O. (1978). Sphingomyelin lipidosis: Niemann–Pick disease. In Stanbury, J.B., Wyngaarden, J.B. and Fredrickson, D.S. (eds.) *The Metabolic Basis of Inherited Disease* 4th Edn., p. 718. (New York: McGraw-Hill)

NINHYDRIN REACTION

This is the reaction of ninhydrin with amino acids to produce a blue colour. It enables the detection and quantitation of amino acids in chromatographic techniques.

NITRAZEPAM (MOGADON)

One of the benzodiazepine group of drugs.

See: **benzodiazepines**

NITRITES

Ingestion of nitrites in the diet has been reported to cause methaemoglobinaemia. Methylene blue or ascorbic acid may be used in treatment.

NITROGEN BALANCE

In a normal person, nitrogen intake equals nitrogen output, 90% of the nitrogen being lost in the urine, mainly as urea. Positive nitrogen balances, where intake exceeds output are found in pregnancy and growing children. Negative nitrogen balances where output exceeds intake are found in malabsorption and protein losing conditions such as burns and nephrotic syndrome. Nitrogen is measured in the urine (and occasionally in the faeces when necessary) by the Kjeldahl technique.

NON-ESTERIFIED FATTY ACIDS (FREE FATTY ACIDS)

In the fasting state, non-esterified fatty acids can supply 50% or more of the body's energy requirements. They are released from adipose tissue and are carried to other tissues where they are further metabolized to supply energy. If there is an excess of non-esterified fatty acids, they are resynthesized into triglycerides and incorporated into pre-β-lipoproteins.

Increases in plasma non-esterified fatty acids are encountered in diabetic ketoacidosis. In this condition there is insufficient α-glycerophosphate (derived from glycolysis) for the free fatty acids (released from adipose tissue) to combine with to re-form triglycerides. As a result of this, the free fatty acids are converted to ketone bodies.

Plasma fatty acid levels can be determined by titration against a known amount of alkali, following their extraction in an organic solvent.

NON-KETOTIC HYPERGLYCINAEMIA

A very rare inborn error of metabolism in which high levels of glycine are found in blood and urine. It is probably due to a deficiency of the enzyme, glycine decarboxylase. Affected individuals suffer from gross mental disturbances and in many cases have epilepsy.

Further reading: Nyhan, W.L. (1978). Nonketotic hyperglycinaemia. In Stanbury, J.B., Wyngaarden, J.B. and

Fredrickson, D.S. (eds.) *The Metabolic Basis of Inherited Disease.* 4th Edn., p. 518. (New York: McGraw-Hill)

NONNE-APELT TEST

A test for the detection of increases in globulin concentration in CSF. It consists of mixing equal volumes of CSF and saturated ammonium sulphate solution. If the CSF globulin content is increased, a turbidity is observed.

NON-PROTEIN NITROGEN (NPN)

These are the substances in blood, other than proteins, which contain nitrogen. They include urea, creatinine, uric acid, amino acids and ammonia. The non-protein nitrogen fraction can be determined by a micro-Kjeldahl method, followed by Nesslerization. In most laboratories, however, blood urea is determined and the other non-protein nitrogen compounds are measured only where clinically indicated, e.g. uric acid.

NORADRENALINE (NOREPINEPHRINE)

A catecholamine hormone, which is synthesized mainly at sympathetic nerve endings and, to a lesser extent, in the adrenal medulla. It is a vasoconstrictor, tending to increase the blood pressure. Increased noradrenaline production occurs in tumours of the adrenal medulla and sympathetic nervous tissue.

See also: **catecholamines**

NOREPINEPHRINE

See: **noradrenaline**

NORMAL RANGE

This is the range of values of a given constituent in a normal population. When establishing a normal range for a particular substance there are a number of factors to be taken into consideration:

(1) It may not be possible to obtain samples from a healthy population. The normal range may have to be established from specimens obtained from inpatients, patients atten-

ding clinics, or other sources. These may not represent the 'true' normal range.

(2) Different age and sex groups may have different normal ranges for the same constituent.

(3) There are a number of different statistical treatments for the data obtained. The most commonly used method is to calculate the range into which 95% of the values fall and call this the normal range. If the values have a symmetrical (normal) distribution, 95% of the individuals are found in the mean ± 2SD range. If the distribution of the values is skewed, the standard deviation cannot be used. However, if the logarithm of the concentration is plotted against the numbers, an approximately normal curve is obtained from which the standard deviation can be calculated.

Further reading: General list of analytical textbooks

NORMETADRENALINE (NORMETANEPHRINE)

A methylated metabolite of noradrenalin.

See: **catecholamines**

NORTRIPTYLINE

A tricyclic antidepressant. It can be detected in body fluids by a number of colorimetric reactions or by its absorption spectrum following its extraction.

NORYMBERSKI PROCEDURE

See: **17-oxogenic steroids**

5′-NUCLEOTIDASE (5′-NT)

A phosphatase enzyme which catalyses the hydrolysis of phosphate from nucleoside 5′-phosphates. Increased serum levels are found in hepatobiliary diseases, together with increased alkaline phosphatase activity. The increase in 5′-nucleotidase activity is usually greater than that of alkaline phosphatase and the higher levels persist for longer. 5′-Nucleotidase measurements can be made in situations where the origin of a raised serum alkaline phosphatase level is sought.

Measurement

Adenosine 5′-monophosphate (AMP) is used as a substrate for 5′-nucleotidase assay. However this substrate can also be hydrolysed by nonspecific phosphatases. Nickel ions inhibit 5′-nucleotidase but not the nonspecific phosphatases. Serum is therefore incubated with AMP, with and without nickel ions and the amounts of inorganic phosphate liberated by the reactions are measured. The difference between the two values corresponds to the activity of 5′ NT.

Further reading: Wilkinson, J.H. (1976). *The Principles and Practice of Diagnostic Enzymology*. (London: Edward Arnold)

O

OAT-CELL BRONCHIAL CARCINOMA

A condition in which ectopic hormone production can occur.

See: **ectopic hormone production**

OBERMAYER'S TEST

A test for the detection of excess indican in urine which consists of adding ferric chloride in acid solution, and chloroform to the sample. A blue colour in the chloroform layer indicates indicanuria.

OCCULT BLOOD IN FAECES

This is the presence of blood in faeces, resulting from bleeding in the alimentary tract. It is called occult because it is not obviously apparent. It can be detected by the peroxidase activity of haemoglobin, in which hydrogen peroxide is used to oxidize dyes such as guaiac. Other peroxidases in the sample (e.g. plant peroxidases) can be destroyed by boiling a suspension of the faeces. The sensitivity of the reaction can also be adjusted in order to avoid false positive results from dietary haemoglobin (i.e. from meat).

Further reading: General list of analytical textbooks

OESTRADIOL

One of the oestrogens secreted by the ovary. For a more detailed account of its role, see **menstrual cycle.** It is metabolized to oestriol and excreted in the urine. Specific oestradiol measurements in plasma and urine can be made by gas chromatography or by radioimmunoassay. It can also be determined, along with other oestrogens, by the Kober reaction. Decreased levels of oestradiol are found in pituitary or ovarian malfunction while increased levels can occur in some ovarian tumours.

See also: **oestrogens**

OESTRIOL

This hormone is synthesized by the placenta and by the fetal liver and adrenals. Its measurement in blood or urine can therefore be used to assess feto-placental function. Large amounts are synthesized and excreted in the urine in late pregnancy. Plasma or urine levels are usually determined at intervals at this stage. A fall in the levels may indicate fetal distress and require obstetric intervention.

Measurement

Oestriol may be measured by radioimmunoassay or gas chromatography in plasma or urine. In many cases it is measured, along with other oestrogens, by colorimetric or fluorimetric methods.

See also: **oestrogens**

OESTROGENS

These are female sex hormones secreted by the ovaries and, in pregnacy, by the placenta. In males, small amounts are secreted by the testes. Small quantities are also synthesized by the adrenals of both sexes. Oestrogens are responsible for the development of the female sexual organs and the secondary sexual characteristics. Quantitatively, the three most important oestrogens are oestrone, oestradiol and oestriol.

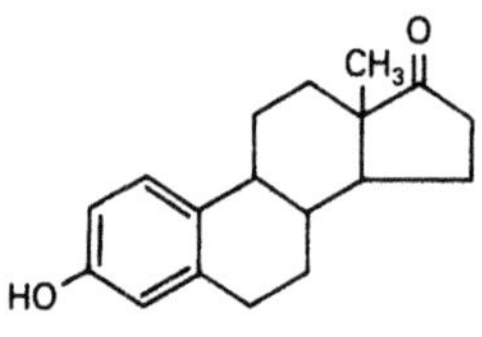

Oestrone

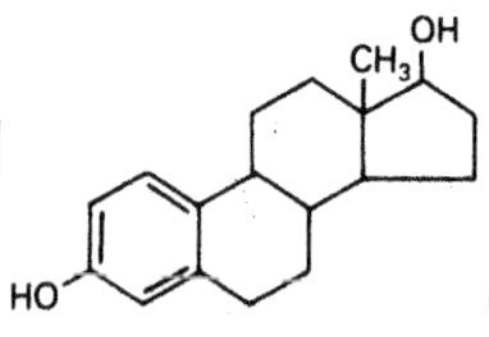

Oestradiol

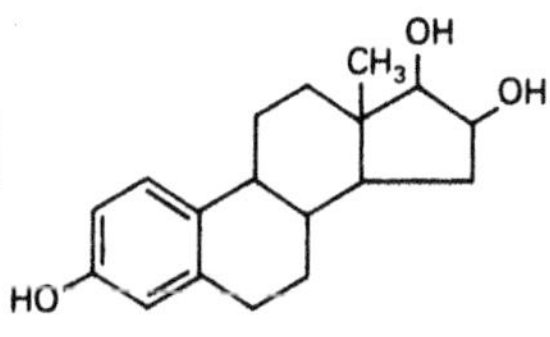

Oestriol

Oestrone and oestradiol are synthesized in the ovary and they are involved in regulation of the menstrual cycle (see **menstrual cycle**). They are excreted as their metabolite, oestriol. In pregnancy, large amounts of oestriol are synthesized by the placenta from precursors manufactured by the fetus. Hence measurement of oestriol in pregnancy serves as an indication of fetoplacental function.

Measurement of total oestrogens

Specific gas-chromatographic and radioimmunoassays are available for the determination of individual oestrogens. However, urinary oestrone, oestriol and oestradiol can be measured collectively as total oestrogens by fluorimetric or colorimetric techniques. Measurements can be performed on the urine of pregnant women and non-pregnant women. In pregnant urine, oestriol forms a considerable part of the total oestrogen fraction.

The determination of total oestrogens in the urine of either pregnant or non-pregnant women consists of an initial hydrolysis of the oestrogen conjugates, extraction of the oestrogens with a number of solvents and then the reaction of the oestrogens with sulphuric acid to form a yellow compound with a green fluorescence (the Kober reaction). Either the absorbance or the fluorescence of this compound can be measured. A number of automated techniques can be used for the determination of total urinary oestrogens.

Significance of the measurement of urinary total oestrogens in non-pregnant women

Decreased excretion of oestrogens can indicate ovarian or pituitary malfunction. Increased excretion of oestrogens is found in some ovarian and testicular tumours.

Significance of the measurement of urinary total oestrogens in pregnant women

Urinary oestrogens are usually determined at intervals in late pregnancy. A fall in the levels during this period may indicate fetal distress and necessitate obstetric action.

Further reading: Hall, R., Anderson, J., Smart, G.A. and Besser, M. (1974). *Fundamentals of Clinical Endocrinology*. 2nd Edn. (London: Pitman Medical Publishing Co.,)

Wilde, C.E. and Oakey, R.E. (1975). Scientific Review No. 3. Biochemical tests for the assessment of fetoplacental function. *Ann. Clin. Biochem.*, **12**, 83

OESTRONE

An oestrogen synthesized by the ovary. It is metabolized to oestriol and excreted in the urine.

See also: **oestrogens**

ONCOFETAL ANTIGENS

Substances produced by a tumour which are normally only produced by embryonic cells, e.g. α-fetoprotein in primary hepatoma, and carcinoembryonic antigen in cancers of the gastrointestinal tract.

See also: **carcinoembryonic antigen, α-fetoprotein**

ONCOTIC PRESSURE (COLLOID OSMOTIC PRESSURE)

The effective osmotic pressure of the blood across capillary walls. The greatest contribution to the colloid osmotic pressure comes from plasma proteins, which, unlike the plasma ions, cannot move through the capillary walls. The oncotic pressure balances the effect of capillary blood pressure which tends to force water into the interstitial spaces.

OPTICAL DENSITY

See: **absorbance**

ORNITHINE

A basic amino acid excreted in the urine in large amounts, along with cystine, arginine and lysine, in the inborn error, cystinuria.

See also: **cystinuria**

ORNITHINE CARBAMYL TRANSFERASE

A urea cycle enzyme found mainly in the liver and kidney. High serum levels are found in conditions where there is hepatocellular damage such as hepatitis. A congenital deficiency of the enzyme has also been described (ornithine carbamyl transferase deficiency).

Further reading: Wilkinson, J.H. (1976). *The Principles and Practice of Diagnostic Enzymology*, (London: Edward Arnold)

ORNITHINE CARBAMYL TRANSFERASE DEFICIENCY

An inborn error of metabolism in which there is a deficiency of this urea cycle enzyme. High levels of ammonia are found in the blood. Vomiting, coma, convulsions and hepatomegaly are associated with the condition.

Further reading: Shih, V.E. (1978). Urea cycle disorders and other hyperammonaemic syndromes. In Stanbury, J.B., Wyngaarden, J.B. and Fredrickson, D.S. (eds.) *The Metabolic Basis of Inherited Disease*. 4th Edn., p. 362. (New York: McGraw-Hill)

OROSOMUCOID

See: **α_1-acid glycoprotein**

OROTIC ACIDURIA

See: **hereditary orotic aciduria**

OSGOOD-HASKIN'S TEST

A test for the detection of Bence Jones proteinuria which consists of adding acetic acid and sodium chloride solution to the urine. Precipitation occurs if Bence-Jones protein is present.

See also: **Bence Jones protein**

OSMOLE

This is the molecular weight in grams of a substance divided by the number of particles or ions into which it dissociates.

OSMOLALITY

A means of expressing osmotic pressure. With blood and urine, it is usually expressed in terms of milliosmoles per kilogram of solvent.

OSMOLARITY

A means of expressing osmotic pressure. With blood and urine, it is usually expressed as milliosmoles per litre of solution.

OSMOTIC PRESSURE

This is the factor which determines the movement of water across cell walls. Hence it is important in fluid homeostatic mechanisms. The movement of water across a semi-permeable membrane depends upon the concentration differences of the dissolved particles on either side of the membrane. Furthermore the osmotic pressure of a solution depends not only on the number of the dissolved particles but also their size. The major contribution to osmotic pressure in normal subjects comes from inorganic ions with only a minor contribution from the proteins.

Situations where measurement of osmotic pressure is useful

Measurements of osmotic pressure (osmometry) are most useful when a direct comparison is made between plasma and urine osmolality. In inappropriate ADH secretion, high urine osmolalities with low plasma osmolalities are found. In diabetes insipidus the opposite is found, i.e. low urine osmolalities with high plasma osmolalities.

Measurement of osmolality

(1) Plasma osmolality can be calculated from a number of formulae if the level of other plasma constituents are known, e.g. sodium, potassium, urea and glucose.

(2) Osmolality can be determined by the measurement of the depression of the freezing point of the sample below that of pure water, a property which is dependent on the number of particles in solution.

(3) Osmolality can be determined in some instruments by measuring the decrease in vapour pressure of the sample below that of pure water, a property which is again dependent on the number of particles in solution.

Further reading: General list of analytical and clinical textbooks

OSTEOMALACIA

A condition in which there is decalcification of bone tissue. It occurs when there is reduced calcium or vitamin D intake, for instance in malabsorption. Osteomalacia can also occur as a result of treatment with anticonvulsant drugs which are thought to interfere with the metabolism of vitamin D. Low serum calcium levels occur along with high serum alkaline phosphatase levels, the latter being due to a secondary increase in osteoblastic activity.

Further reading: General list of clinical textbooks

OSTEOPOROSIS

A condition which clinically resembles osteomalacia. It differs from osteomalacia, however, in that it is due to a primary loss of the proteinaceous bone matrix with a resultant secondary loss of bone calcium. It can occur as part of the general process of ageing, or in conditions such as hyperthyroidism, acromegaly, Cushing's syndrome (because of the effects of thyroxine, growth hormone and cortisol on protein metabolism) and malabsorption (due to deficient amino acid absorption). Serum calcium, phosphate and alkaline phosphatase levels are normal, unlike osteomalacia.

Further reading: General list of clinical textbooks

OUCHTERLONY TECHNIQUE

See: **double diffusion test**

OVARIES

Female reproductive organs which secrete a number of hormones including oestrogens (oestradiol and oestrone), progesterone and androgens (the main androgen being androstenedione). For a more detailed account of the hormonal changes in the ovaries see **menstrual cycle**,

Diagnosis of ovarian disorders

(1) Hypogonadism may be due to primary ovarian failure or secondary to pituitary hypofunction. In primary gonadal failure, oestrogen levels are low while the plasma levels of FSH and LH may be normal or raised. In secondary hypogonadism, oestrogens, FSH and LH levels are low. The clomiphene stimulation test (qv) and the LH/FSH releasing hormone stimulation test (qv) can help in distinguishing between primary and secondary hypogonadism.

(2) Plasma progesterone (or its metabolite pregnanediol) can be measured during the second half of the menstrual cycle. A normal level indicates that ovulation has taken place.

(3) Many ovarian tumours secrete oestrogens. In some ovarian tumours, 17-oxosteroid excretion and plasma testosterone levels may be elevated.

(4) HCG may be excreted in cases of ovarian teratoma.

(5) In patients with polycystic ovary syndrome, 17-oxosteroid excretion and testosterone levels may be elevated.

Further reading: Hall, R., Anderson, J., Smart, G.A. and Besser, M. (1974). *Fundamentals of Clinical Endocrinology*. 2nd Edn., (London: Pitman Medical Publishing Co.)

OXALIC ACID

A carboxylic acid, normally excreted in the urine in small amounts. It is a constituent of many urinary tract stones. High levels of oxalic acid are excreted in the urine in the rare inborn error of metabolism, primary hyperoxaluria. In this disorder renal stones composed of oxalate are formed and death results from progressive renal failure. The increase in the urinary excretion of oxalic acid appears to be derived from glycine as a result of deficient glyoxylic acid–glycine transamination.

Further reading: Williams, H.E. and Smith, L.H. Jr. (1978). Primary hyperoxaluria. In Stanbury, J.B., Wyngaarden J.B. and Fredrickson, D.S. (eds.) *The Metabolic Basis of Inherited Disease*. 4th Edn., p. 182. (New York: McGraw-Hill)

17-OXOGENIC STEROIDS (17-KETOGENIC STEROIDS)

These are urinary corticosteroids and their metabolites. Among the substances measured as 17-oxogenic steroids are cortisol, cortisone and 11-deoxycortisol with their tetrahydro derivatives, and cortol and cortolone. The term 17-oxogenic steroids is often used synonymously with the term 17-hydroxysteriods and for most purposes means the same thing. However, because of the differences in the methods by which these two groups are determined, there is a slight difference between the two (see below and also **17-hydroxycorticosteroids**).

Low values are found in adrenal hypofunction while high values are found in Cushing's syndrome and congenital adrenal hyperplasia.

Measurement

The term 17-oxogenic steroids is used because, in their determination, the corticosteroids and their metabolites are oxidized to 17-oxosteroids which can then be determined by the Zimmermann reaction. This is the Norymberski procedure. The initial step in their determination consists of borohydride reduction, by which the corticosteroids and their metabolites are converted to their dihydroxy derivatives. At the same time, the 17-oxo steroids present are reduced to secondary alcohols which are not affected by bismuthate or periodate (see below) and do not give the Zimmermann reaction. The next step consists of bismuthate or periodate oxidation, by which the dihydroxy derivatives are converted to 17-oxosteroids. These are then estimated by their reaction with *m*-dinitrobenzene in alcoholic alkali to give a purple compound (the Zimmermann reaction).

See also: **cortisol, 17-hydroxycorticosteroids**

Further reading: General list of analytical and clinical textbooks

17-OXOSTEROIDS (17-KETOSTEROIDS)

17-Oxosteroids are androgens and their metabolites. They are secreted by the adrenals and, to a lesser extent, by the testes and ovaries. In females, most of the 17-oxosteroids originate from the adrenals, while in males, about 15% of the total is derived from metabolites of the testicular hormone, testosterone. Among the compounds measured as 17-oxosteroids are dehydroepiandrosterone, aetiocholanelone derivatives and androsterone derivatives. In most cases, 17-oxosteroid measurement is used as an index of androgen production by the adrenals, high values being found in adrenal hyperplasia and adrenal carcinoma, low values being found in adrenal hypofunction. They can also be used to assess gonadal function, low values occuring in hypogonadism while high values can be found in testicular tumours.

Measurement

17-Oxosteroids can be extracted from urine using suitable solvents and estimated by their reaction with *m*-dinitrobenzene in alcoholic alkali to give a purple colour (the Zimmermann reaction).

Further reading: General list of analytical and clinical textbooks

OXYGEN

Oxygen in blood is carried loosely bound to haemoglobin. Only a small proportion exists in simple solution. The oxygen parameter most commonly measured is the PO_2, the partial pressure of this dissolved oxygen. This gives an indication of the oxygen status of the body since the dissolved oxygen is in equilibrium with the oxygen bound to haemoglobin. Other parameters of oxygen status can be measured however. See **oxygen content, oxygen capacity, oxygen combining power, oxygen saturation and P_{50}**.

Decreased arterial PO_2

Low PO_2s are found in chronic obstructive lung diseases and cardiac failure. It should be remembered that hypoxia may still occur in the tissues in conditions such as anaemia, when the PO_2 is normal but the oxygen capacity of the blood is decreased because of the low haemoglobin level.

Determination of PO_2

PO_2 can be determined by a means of a PO_2 electrode (Clark electrode). In this technique, oxygen diffuses from the blood sample across a gas-permeable membrane into an electrochemical system which consists of a platinum cathode and a silver/silver chloride anode. Reduction of the oxygen occurs at the cathode, resulting in the generation of a current which can be measured, the current being directly proportional to the PO_2. This is an example of an amperometric technique

Further reading: General list of analytical and clinical text-books

11-OXYGENATION INDEX

A determination used in the diagnosis of congenital adrena hyperplasia. The final step in the biosynthesis of cortisol is ; hydroxylation of the steroid molecule at position 11. When ther is a metabolic block in the pathway, as occurs in congenita adrenal hyperplasia, there will be a decreased proportion of 17 oxogenic steroids with a hydroxyl group at postion 11. The rati of the steroids without an 11-hydroxyl group to those with a 11-hydroxyl group is termed the 11-oxygenation index, and i congenital adrenal hyperplasia it will be higher than normal.

Measurement

The steroids in the urine are reduced by borohydride, the oxidized by periodate (see 17-oxogenic steroids). The 11-deox and 11-hydroxy derivatives are separated by differential solve extraction using organic solvents and the Zimmermann reaction is applied to both fractions.

See also: **congenital adrenal hyperplasia**

OXYGEN CAPACITY OF THE BLOOD

The amount of oxygen which the blood contains when it is fully oxygenated. It represent both oxygen in physical solution and oxygen bound to haemoglobin. If the oxygen in physical solution is subtracted from the oxygen capacity, the oxygen combining capacity is obtained.

OXYGEN COMBINING POWER OF THE BLOOD

The amount of oxygen combined with haemoglobin when the haemoglobin is fully saturated.

OXYGEN CONTENT OF THE BLOOD

This is the total oxygen present in the blood, i.e. the sum of the dissolved oxygen (the PO_2) and the oxygen bound to haemoglobin.

OXYGEN SATURATION OF THE BLOOD

This is the percentage of the oxygen combining power contributed by the oxygen actually bound to haemoglobin. Normal arterial oxygen saturation is about 95–98%.

OXYGEN UNSATURATION OF THE BLOOD

This is obtained by subtracting the oxygen saturation from 100, i.e. it is the opposite of oxygen saturation.

OXYHAEMOGLOBIN

Haemoglobin in the oxygenated form.

OXYTOCIN (PITOCIN)

A polypeptide hormone secreted by the posterior pituitary gland which stimulates uterine contractions during parturition. It also stimulates milk secretion.

OXYTOCINASE

See: **cystine aminopeptidase**

P

***P*$_{50}$**

This is the PO_2 at half saturation of haemoglobin. The P_{50} is a measurement of the affinity of haemoglobin for oxygen, a low P_{50} indicating a high affinity and *vice versa*. Among the factors which influence the P_{50} are pH, PCO_2, temperature and 2,3-diphosphoglycerate levels.

See also: **Bohr effect, 2,3-diphosphoglycerate**

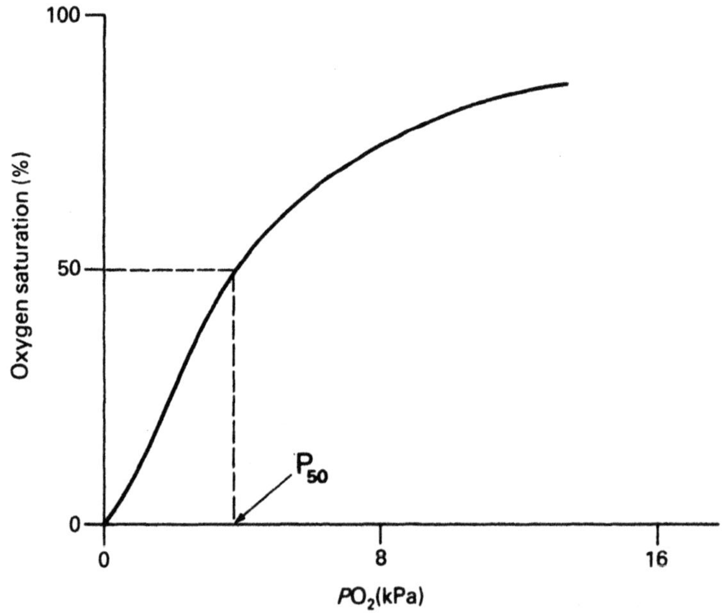

PAGET'S DISEASE OF BONE

A bone disease in which there is over-activity of osteoblasts as they try to compensate for the bone which is resorbed by the uncontrolled action of osteoclasts. This leads to dense bone formation with areas of rarefaction. Serum calcium and phosphate levels are not usually affected but the alkaline phosphatase level can be very high.

PALMITIC ACID

Amniotic fluid palmitic acid can be measured as an indication of fetal lung maturity as an alternative to lecithin–sphingomyelin ratios (q.v.). This is because palmitic acid is a component of the lecithin molecule. High amniotic fluid palmitic acid levels indicate fetal lung maturity and enable early delivery of the infant to be contemplated. Palmitic acid can be estimated by gas–liquid chromatography after hydrolysis of the lecithin to give the free fatty acid.

PANCREAS

A gland which has both an endocrine function (the secretion of insulin and glucagon) and an exocrine function (the secretion of an alkaline fluid containing digestive enzymes into the duodenum). The digestive enzymes which are most frequently measured as an index of pancreatic function are trypsin, amylase and lipase.

Further reading: Wormsley, K.G. (1977). Pancreatic exocrine physiology. *Br. J. Hosp. Med.,* **18**, 518

PANCREATIC STIMULATION TESTS

These are dynamic tests which assess the ability of the pancreas to secrete an alkaline fluid rich in digestive enzymes.

See also: **Lundh test, secretin stimulation test, secretin-cholecystokininpancreozymin stimulation test**

PANCREATITIS

Acute pancreatitis occurs most commonly as a result of obstruction of the pancreatic duct. Diseases of the biliary tract and alcoholism are factors which can lead to this condition. The symptoms of acute pancreatitis include abdominal pain and shock and these are thought to be due to pancreatic enzymes in the abdominal cavity. The biochemical features of acute pancreatitis include raised serum levels of pancreatic enzymes (amylase being the one most commonly measured) and hypocalcaemia, caused by the reaction of calcium with fatty acids released from fats by the action of lipase. Repeated attacks of acute pancreatitis can result in the chronic form of the disease in which malabsorption is a feature. Chronic pancreatitis is more difficult to diagnose because of the variations in serum amylase levels from day to day.

Further reading: Hermon-Taylor, J. (1977). An aetiological and therapeutic review of acute pancreatitis. *Br. J. Hosp. Med.,* **18**, 546

Mallinson, C. (1977). Chronic pancreatitis. *Br. J. Hosp. Med.,* **18**, 553

PANCREOZYMIN

See: **cholecystokinin-pancreozymin**

PANDY'S TEST

A test for the detection of high levels of globulins in cerebrospinal fluid. It consists of adding two drops of the CSF to saturated phenol solution. A turbidity results if the globulin levels are high.

PANHYPOPITUITARISM

See: **hypopituitarism**

PANTOTHENIC ACID

A water-soluble vitamin of the B group. It is a component of coenzyme A which participates in many reactions in intermediary metabolism.

PAPER CHROMATOGRAPHY

A form of partition chromatography (q.v.)

PARACETAMOL (*p*-ACETYLAMINOPHENOL, ACETAMINOPHEN)

An analgesic drug which is increasingly encountered in overdose cases. In such patients, it may have fatal consequences, as a result of massive liver damage, several days after ingestion. Paracetamol overdoses can be treated with drugs such as cysteamine. As these drugs have toxic side effects, it is desirable to know if the patient has taken sufficient paracetamol tablets to warrant treatment. Hence it may be necessary to determine the plasma concentration of paracetamol. This should ideally be performed within 12 hours of ingestion as paracetamol is rapidly cleared from the body.

The plasma level of paracetamol can be determined by gas–liquid chromatography, by its UV absorption spectrum following extraction, and by its colorimetric reaction with nitrous acid to form a yellow coloured nitrophenol (the Glynn and Kendal method). Paracetamol can be detected in urine by a screening test which consists of its hydrolysis to *p*-aminophenol, followed by its reaction with *o*-cresol and ammonia to form a blue indophenol.

Further reading: Wiener, K. (1978). A review of methods for plasma paracetamol estimation. *Ann. Clin. Biochem.*, **15**, 187

Editorial. (1975). Paracetamol (acetaminophen) and the liver. *Br. Med. J.*, **1**, 536

PARAPROTEIN

An abnormal immunoglobulin or part of an immunoglobulin, resulting from the proliferation of a single type of immunocyte (plasma cell or lymphocyte). It can appear as an extra band on serum protein electrophoresis. The abnormal immunoglobulin is of a single class and type. The diseases which result in the production of a paraprotein are called monoclonal gammopathies, the most common being myeloma (q.v.) and Waldenström's macroglobulinaemia (q.v.) although paraproteinaemia can also be found in benign conditions. Occasionally abnormal bands may be found on serum protein electrophoresis which are not immunoglobulins. Such 'pseudo' paraprotein bands can be due to haemoglobin (due to haemolysis), C-reactive protein (in inflammatory diseases), fibrinogen (if plasma has been used instead of serum) or because the serum is old or uraemic.

Further reading: Martin, N.H. (1970). The paraproteinaemias. *Br. J. Hosp. Med.*, **3**, 662

Kohn, J. (1973). The laboratory investigation of paraproteinaemia. *Recent Adv. Clin. Pathol.*, **6**, 363

PARAQUAT

A highly toxic week killer which can result in death by pulmonary fibrosis. It can be detected in urine by its blue colour following its reduction with alkaline dithionite.

Further reading: Editorial. (1976). Paraquat poisoning. *Lancet*, **1**, 1057

PARATHYROID HORMONE (PARATHORMONE, PTH)

A polypeptide hormone secreted by the parathyroid glands in response to low circulating ionized calcium levels. It has three actions which tend to raise the level of serum ionized calcium:

(1) It acts directly on osteoclasts, releasing bone salts into the circulation.

(2) It decreases the reabsorption of phosphate by the rena tubules. This decreases the plasma phosphate level and thi causes the release of phosphate and therefore calcium fron bone.

(3) It may also stimulate the hydroxylation of vitamin D in th liver and kidney.

Radioimmunoassay can be used to estimate the level c the hormone. For an account of the pathology of PTH *se* **hyperparathyroidism and hypoparathyroidism.**

Further reading: Tomlinson, S. and O'Riordan, J.L.H. (1978) The parathyroids. *Br. J. Hosp. Med.*, **19**, 40

PARIETAL CELL ANTIBODIES

Autoantibodies which act against gastric parietal cells. They ca be detected in the serum of many patients with perniciou anaemia.

PARTITION CHROMATOGRAPHY

A form of chromatography by which a mixture of molecules ca be separated on the basis of their differential distributio between two phases, one being stationary and the other mobil The stationary phase is a liquid bound onto a solid support. Th solid support may be paper (as in paper chromatography), or cellulose or silica (as in thin-layer or column chromatography), or a diatomaceous earth (as in gas–liquid chromatography). In paper, column and thin-layer chromatography, the stationary phase is usually aqueous, while in gas–liquid chromatography it is generally a high temperature organic liquid, such as silicone or polyethylene glycol. The mobile phase is usually an organic liquid except in the case of gas–liquid chromatography where it is an inert gas. Thin-layer and paper chromatography will be discussed below. See elsewhere for a discussion of gas–liquid chromatography.

The mixture to be resolved is spotted onto a paper or thin-layer plate which is placed in a tank containing a shallow reservoir of the solvent. The liquid, which is usually a mixture of water and an organic liquid, rises up the support medium and the molecules to be resolved partition themselves between the stationary and mobile phases. Polar compounds, being more soluble in the polar stationary phase, therefore migrate more slowly than non-polar compounds which are more soluble in the mobile phase. This differential solubility results in their eventual resolution. The separated components can be visualized by spraying with a suitable reagent.

PCO_2

See: **carbon dioxide**

PELLAGRA

The condition which can result from a deficiency of nicotinamide (q.v.)

PENICILLAMINE

A chelating agent used in the treatment of a variety of conditions. These include lead poisoning and Wilson's disease where it helps remove lead and copper respectively. It is also used in the treatment of cystinuria.

PENTAGASTRIN STIMULATION TEST

Pentagastrin is a pentapeptide which represents the physiologically active portion of the gastrin molecule. It stimulates the gastric parietal cells to secrete acid and therefore can be used to assess gastric function. In the test itself, pentagastrin is injected intramuscularly and the gastric juice aspirated and analysed for acid content. Low acid outputs occur in cases of gastric carcinoma and pernicious anaemia.

PENTOBARBITONE

An intermediate acting barbiturate

See: **barbiturates**

PENTOSURIA

Essential pentosuria is a benign inborn error of metabolism found mainly in Jews. It is due to a deficiency of xylitol dehydrogenase, an enzyme involved in the oxidation of glucuronic acid to pentoses. This results in increasing excretion of L-xylulose in the urine. The condition has a recessive mode of inheritance.

Further reading: Hiatt, H.H. (1978). Pentosuria. In Stanbury, J.B., Wyngaarden, J.B. and Fredrickson, D.S. (eds.) *The Metabolic Basis of Inherited Disease*. 4th Edn., p. 110. (New York: McGraw-Hill)

PEPSIN, PEPSINOGEN

Pepsin is a gastric enzyme which hydrolyses proteins. It is secreted by the stomach as its inactive precursor, pepsinogen. This is converted to pepsin by the action of gastric acid.

See also: **uropepsin, uropepsinogen**

PERCENTILE

The value below which a certain proportion of a number of observations fall. For instance the 10th percentile means that 10% of the observations are below this value.

PERNICIOUS ANAEMIA

A macrocytic anaemia which results from failure to absorb vitamin B_{12}. In true pernicious anaemia it is due to failure of the stomach to produce adequate amounts of intrinsic factor. Intrinsic factor antibodies and/or gastric parietal cell antibodies can be demonstrated in many patients with pernicious anaemia.

PETTENKOFFER'S TEST

A test for the detection of excess bile salts in urine based on their reaction with sucrose and concentrated sulphuric acid to give a red colour.

pH

A measure of hydrogen ion activity. It is given by the formula:

$$pH = -\log_{10} [H+]$$

where [H+] is the hydrogen ion concentration.

From the Henderson–Hasselbalch equation, the pH of blood can be related to the bicarbonate and carbon dioxide concentrations as follows:

$$pH \propto \frac{[\text{bicarbonate}]}{[CO^2]}$$

Bicarbonate can be considered as the metabolic component of the acid–base picture and carbon dioxide as the respiratory component. Low blood pHs are found in acidotic conditions, while high blood pHs are found alkalotic conditions.

Measurement

The pH of biological and other fluids is usually measured by a glass electrode. This is selective to hydrogen ions and can be considered as an ion-selective electrode (q.v.)

See also: **acid–base balance, acidosis, alkalosis, bicarbonate and carbon dioxide**

PHAEOCHROMOCYTOMA

A catecholamine secreting tumour of chromaffin tissue occuring mainly in adults. The majority of these tumours are located in the adrenal medulla. It can be diagnosed by measurement of catecholamines or their metabolites such as HMMA.

See also: **catecholamines**

PHENACETIN

An analgesic drug which is metabolized to its pharmacologically active metabolite, paracetamol. However, it is now infrequently used because of its toxic effects on the kidney.

PHENISTIX

A dipstick test, manufactured by Ames for the detection of phenylpyruvic acid in urine, based on its reaction with ferric ions to yield a grey-green colour. It can therefore be used as a screening test for phenylketonuria.

Further reading: Kutter, D. (1977). *Rapid Clinical Diagnostic Tests*. (Munich–Vienna–Baltimore: Urban and Schwarzenberg)

PHENOBARBITONE

A long-acting barbiturate.

See: **barbiturates**

PHENOL RED TEST

See: **phenolsulphonephthalein test**

PHENOLSULPHONEPHTHALEIN TEST (PHENOL RED TEST)

Phenolsulphonephthalein (PSP) is a dye which is excreted in the urine, mainly as a result of renal tubular secretion. Measurement of its excretion in the urine after its intravenous injection therefore gives an estimate of the renal tubular secretory capacity. The dye can be estimated in urine by adding alkali which converts the dye into its coloured form.

PHENOTHIAZINES

A class of drugs which are used as tranquillizers or antihistamines. They can be detected in urine by a variety of colou reactions, e.g. with FPN (ferric chloride–perchloric acid–nitric acid) reagent.

Further reading: Yeoman, W.B. (1971). Toxicological analysis in the clinical chemistry laboratory. *Ann. Clin. Biochem.*, **8**, 93

Meade, B.W. *et al.* (1972). Technical Bulletin No. 24. Simple tests to detect poisons. *Ann. Clin. Biochem.*, **9**, 35

PHENYLALANINE

An aromatic amino acid found in increased levels in the blood and urine in the inborn error of metabolism, phenylketonuria. It can be estimated by microbiological assay (the Guthrie test, q.v.) or by its reaction with ninhydrin to form a fluorescent compound. This fluorescence is enhanced by the addition of the dipeptide, L-leucylalanine.

PHENYLALANINE TOLERANCE TEST

A test that can be used for the detection of carriers (heterozygotes) of phenylketonuria. It consists of giving an oral load of phenylalanine and measuring the subsequent blood levels. In normals, the blood phenylalanine levels rise and then return to near their original value. In heterozygotes, the level remains raised for much longer.

PHENYLKETONURIA

An inborn error of metabolism in which there is a deficiency of phenylalanine hydroxylase, the enzyme which converts phenylalanine to tyrosine. As a result, phenylalanine accumulates in the body and is converted to a number of derivatives such as phenylpyruvic acid, phenyllactic acid and phenylacetic acid which are excreted in the urine. If the condition is untreated, mental retardation may result. It has an autosomal recessive mode of inheritance.

The condition may be diagnosed by a number of different tests:

(1) The phenylalanine concentration may be measured in blood either by chemical estimation or by microbiological assay (the Guthrie test, q.v.). The latter technique can be used as a screening procedure.

(2) Phenylpyruvic acid may be detected in the urine by Phenistix or by ferric chloride.

(3) A deficiency of the enzyme can be demonstrated in a liver biopsy.

(4) A phenylalanine tolerance test (q.v.) can be used for the detection of heterozygotes.

Diets low in phenylalanine are used to treat the condition. Phenylalanine levels must be monitored at regular intervals in affected individuals.

Further reading: Tourian, A.Y. and Sidbury, J.B. (1978). Phenylketonuria. In Stanbury, J.B., Wyngaarden, J.B. and Fredrickson, D.S. (eds.) *The Metabolic Basis of Inherited Disease*. 4th Edn., p. 240. (New York: McGraw-Hill)
Editorial. (1979). New varieties of PKU. *Lancet*, **1**, 304

PHENYTOIN

An anticonvulsant drug. Among its undesirable side-effects is that of osteomalacia. This is thought to be due to the stimulation by the drug of hepatic enzymes which inactivate vitamin D resulting in decreased calcium absorption. Phenytoin also interferes with folate metabolism and this may result in a megaloblastic anaemia.

Further reading: Editorial. (1976). Anticonvulsant osteomalacia. *Br. Med. J.*, **2**, 1340

PHOSPHATASE

See: **acid phosphatase and alkaline phosphatase**

PHOSPHATE

The majority of the phosphorus in the body occurs as calcium phosphate salts in teeth and bones. The remainder occurs in phospholipids, nucleic acids, organic phosphate esters and inorganic phosphate. The inorganic phosphate in the blood contributes towards its buffering capacity.

The metabolism of phosphate is closely linked to that of calcium. The factors which regulate calcium metabolism also influence phosphate metabolism. Among these factors are:

(1) *Parathyroid hormone.* This acts directly on osteoclasts, releasing calcium and phosphate into the blood. It also decreases the tubular reabsorption of phosphate. This tends to decrease the serum phosphate level which in turn increases the release of phosphate salts from bone.

(2) *Calcitonin.* This decreases osteoclastic activity. It also has a phosphaturic effect.

(3) *Vitamin D.* In addition to increasing calcium and phosphate absorption from the gut, it too has a phosphaturic effect.

Other factors not related to bone metabolism also influence the plasma phosphate level. The phosphate level is lowered after a carbohydrate meal. This is because carbohydrate metabolism promotes the entry of phosphate into cells where they are esterified into organic esters. It is therefore the custom in many hospitals to measure the fasting serum phosphate. This in general varies inversely with the calcium concentration, although there are exceptions to this.

Causes of a high serum phosphate level

(1) Hypoparathyroidism and pseudohypoparathyroidism.
(2) Vitamin D overdosage.
(3) Renal failure.
(4) Acromegaly.

Causes of a low serum phosphate level

(1) Hyperparathyroidism.
(2) Insulin therapy–this causes phosphate to enter cells.
(3) Osteomalacia
(4) Hypopituitarism

Measurement

Blood samples for inorganic phosphate estimation should be rapidly separated as phosphate diffuses from the cells following the enzymic hydrolysis of phosphate esters. It is therefore advantageous to collect such blood specimens in containers having a fluoride preservative. This inhibits the enzymatic hydrolysis. Several methods are available for inorganic phosphate estimation.

(1) The most commonly used techniques are based on the reaction of phosphate with a molybdate reagent to form

phosphomolybdate. This is then reduced by a suitable reducing agent to give a molybdenum blue which is estimated colorimetrically. Among the reducing agents which can be used are aminonaphtholsulphonic acid, *p*-methylaminophenol (metol) and stannous chloride. These methods have been adapted for use on continuous flow instruments and also for use on discrete analysers, often without the need for protein precipitation.

(2) Phosphomolybdate absorbs light at 340 nm and therefore some methods can be based on the direct measurement of this compound by ultraviolet spectrophotometry. This dispenses with the reduction stage.

(3) Other methods of phosphate estimation are based on its reaction with molybdate and vanadate to give yellow phosphomolybdovanadate which can be estimated colorimetrically.

(4) Phosphate can be measured by the reaction of phosphomolybdate with the dye malachite green. This results in a shift in the spectral characteristics of the dye which can be measured colorimetrically.

Further reading: General list of analytical and clinical textbooks

PHOSPHATE EXCRETION INDEX

The ratio of the phosphate clearance (C_p) to the creatinine clearance (C_{cr}) gives an indication of the proportion of phosphate filtered at the glomerulus which has been reabsorbed by the renal tubules (a process in which parathyroid hormone is involved). In order to allow for fluctuations in the level of serum phosphate, the phosphate excretion index (PEI) was devised and this is given by the formula:

$$\mathrm{PEI} = \frac{C_p}{C_{cr}} - 0.05\,(x-1)$$

where x is the serum phosphate concentration in mg per 100 ml.

Normal values for the PEI range between −0.09 and +0.09. The PEI is raised in hyperparathyroidism and lowered in hypoparathyroidism.

PHOSPHOLIPIDS

These are complex lipids containing phosphate and a nitrogenous base. They are found in plasma lipoproteins, in biological membranes and in bile where they help to maintain cholesterol in solution. Phospholipids can be estimated by digestion of a lipid extract followed by measurement of the inorganic phosphate liberated.

PHOSPHORESCENCE

The persistent emission of light by molecules, following their excitation by light and which continues after the exciting radiation has been removed (unlike fluorescence when the molecules emit light only in the presence of the exciting radiation).

Further reading: Rubin, M. (1970). Fluorimetry and phosphorimetry in clinical chemistry. In Bodansky, O. and Stewart, C.P. (eds.) *Advances in Clinical Chemistry*. Vol. 13, p. 163. (New York: Academic Press)

PHOTOMETRY

See: **spectrophotometry**

PHOTOMULTIPLIER TUBE

A device for the measurement of light intensity. They can be found in spectrophotometers and also in scintillation counters for the measurement of radioactivity.

Photomultiplier tubes have a photosensitive cathode to which is applied an external voltage. Light striking the cathode displaces electrons, the number displaced being proportional to the intensity of the light. The electrons go through a number of other stages where voltages are applied and, at each stage, more electrons are displaced. In this way each electron produces a cascade of other electrons by means of the photomultiplication stages and this can be measured electronically. The main advantage of photomultiplier tubes is their rapid response times.

Further reading: General list of analytical textbooks

PHYTANIC ACID STORAGE DISEASE

See: **Refsum's disease**

PHYTATE

A compound found in certain plant foods, especially chapattis (a food used by Asian communities). It forms insoluble salts with calcium in the gut and thereby inhibits its absorption. Hypocalcaemia with osteomalacia can result.

PILOCARPINE

A drug that can be introduced into the skin by iontophoresis. It stimulates sweat secretion during the sweat test for the diagnosis of cystic fibrosis.

See also: **sweat test**

PITOCIN

See: **oxytocin**

PITRESSIN

See: **antidiuretic hormone**

PITRESSIN TEST

See: **water deprivation test**

PITUITARY GLAND

Anatomically, the pituitary consists of two parts.

(1) *Anterior pituitary* (adenohypophysis)
This secretes the following hormones:
Adrenocorticotrophic hormone
Follicle stimulating hormone
Growth hormone
Luteinizing hormone
Melanocyte stimulating hormone
Prolactin
Thyroid stimulating hormone

Secretion of these hormones is under the control of hypothalamic release and inhibiting factors (q.v.)

(2) *Posterior pituitary* (neurohypophysis)
This secretes:

Antidiuretic hormone
Oxytocin.

See: separate entries for these hormones.

Further reading: Hall, R., Anderson, J., Smart, G.A. and Besser, M. (1974). *Fundamentals of Clinical Endocrinology*. 2nd Edn., (London: Pitman Medical Publishing Co.)

Supplement. Hypothalamic and Pituitary Hormones. *J. Clin. Path.* April 1979

PLACENTAL FUNCTION

See: **pregnancy**

PLACENTAL LACTOGEN

See: **human placental lactogen**

PLASMA

The fluid which remains after red cells have been removed from blood in which clotting has been prevented by the use of an anticoagulant.

PLASMACYTOMA

A malignant proliferation of a single type of plasma cell usually in the bone marrow in which case it is known as a myeloma (qv), but very occasionally occuring in other tissues (soft-tissue plasmacytoma, qv).

PLASMIN

An enzyme present in blood which lyses fibrin and fibrinogen. Activators such as trypsin convert the inactive precursor plasminogen into plasmin.

PLEURAL FLUID

The only commonly measured substance in pleural fluid is protein. High levels (greater than 30 g/l) are thought to occur if the fluid is an exudate (fluid arising from neoplasms or infections). Levels below 30 g/l are thought to arise from transudates. This may occur for example in cases of congestive cardiac failure.

Further reading: Hickman, J.A. (1975). The analysis of pleural fluid. *Br. J. Hosp. Med.*, **14**, 624

PO_2

See: **oxygen**

POLAROGRAPHY

A technique which consists of the measurement of both the current flowing through an electrochemical cell and the electrical potential between the two electrodes while this is increased by an external electrical source.

POLYCLONAL GAMMOPATHY

Diseases in which the γ-globulin fraction of the serum proteins shows a diffuse increase on electrophoresis due to the greater than normal production of many types of immunoglobulin molecules (*cf* a monoclonal gammopathy when only one class and type of immunoglobulin is produced and this is seen as an extra band on serum protein electrophoresis). Chronic infections, liver cirrhosis and inflammatory disorders such as rheumatoid arthritis are examples of polyclonal gammopathies.

POLYCYSTIC OVARY SYNDROME

See: **Stein–Leventhal syndrome**

POLYVINYLPYRROLIDONE (PVP)

Radioactively labelled PVP can be used in the diagnosis of protein-losing enteropathies, since it has a molecular weight similar to albumin. It is injected intravenously and its loss is measured in the faeces.

POMPE'S DISEASE

One of the glycogen storage diseases (qv).

PORPHOBILINOGEN

An intermediate in the synthesis of haem. Increased urinary excretion of porphobilinogen occurs in many of the porphyrias. It can be estimated by its condensation with *p*-dimethylaminobenzaldehyde (Ehrlich's reagent) to give a red colour.

PORPHYRIAS

A group of disorders in which there is disturbance of porphyrin biosynthesis resulting in increased blood and tissue levels of porphyrins or their precursors. Most types of porphyria are hereditary while one (symptomatic cutaneous hepatic porphyria) is acquired. The porphyrias can be classified into several groups depending upon whether the basic abnormality is in the liver or the erythropoietic system or both:

(1) Hepatic porphyrias:

Acute intermittent porphyria
Hereditary coproporphyria
Porphyria variegata
Symptomatic cutaneous hepatic porphyria

(2) An erythropoietic porphyria: congenital erythropoietic porphyria.

(3) An erythrohepatic porphyria: protoporphyria.

See separate entries for each of these conditions.

Among the tests which can be used in the investigation of the porphyrias are measurement of porphyrins, porphobilinogen or δ-aminolaevulinic acid in faeces and urine.

Further reading: Meyer, U.A. and Schmid, R. (1978). The porphyrias. In Stanbury, J.B., Wyngaarden, J.B. and Fredrickson, D.S. (eds.) *The Metabolic Basis of Inherited Disease*. 4th Edn., p. 1166 (New York: McGraw-Hill)

Elder, G.H., Gray, C.H. and Nicholson, D.C. (1972). The porphyrias: a review. *J. Clin. Pathol.*, **25**, 1013

PORPHYRIA VARIEGATA

An inherited type of hepatic porphyria which is relatively common among white South Africans. It is similar to acute intermittent porphyria but differs from it in that skin lesions are a feature. Like acute intermittent porphyria, δ-ALA, porphobilinogen and porphyrins are excreted in the urine during acute attacks. The main biochemical abnormality is however the excessive secretion of porphyrins in the faeces. It has a dominant mode of inheritance.

Further reading: Meyer, U.A. and Schmid, R. (1978). The porphyrias. In Stanbury, J.B., Wyngaarden, J.B. and Fredrickson, D.S. (eds.) *The Metabolic Basis of Inherited Disease.* 4th Edn., p. 1166 (New York: McGraw-Hill)

PORPHYRINS

A group of tetrapyrrole compounds formed during the biosynthesis of haem, cytochromes and other compounds. They include coproporphyrins, uroporphyrins and protoporphyrins.

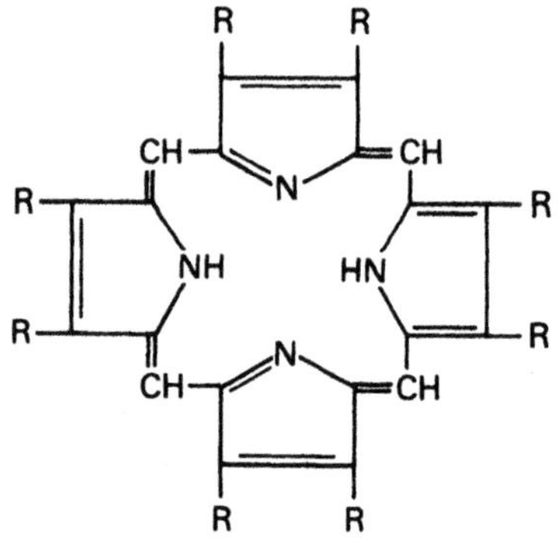

Side groups

Uroporphyrins	Coproporphyrins	Protoporphyrins
R = Acetate and propionate	R = Methyl and propionate	R = Methyl propionate and vinyl

Type I and III isomers differ in the order of the side groups on one of the rings

Abnormal amounts of porphyrins are excreted in the group of disorders known as porphyrias. In other conditions, such as lead poisoning, urinary porphyrin excretion can also occur.

Porphyrins can be detected in urine and faeces by extraction into an organic solvent which is then examined for fluorescence. Coproporphyrins, protoporphyrins and uroporphyrins can also be measured separately in urine and faeces. This involves their differential extraction, using a number of different solvent systems. followed by measurement of their absorbance using a spectrophotometer. Various chromatographic procedures can also be used for their separation.

Further reading: General list of analytical textbooks

PORTER–SILBER REACTION

The reaction of certain corticosteroids with phenylhydrazine in the presence of alcohol and sulphuric acid to give a yellow colour. Steroids which give this reaction have a dihydroxy-acetone side chain and they include such compounds as cortisol, 11-deoxycortisol, cortisone and their tetrahydro derivatives. Collectively they are known as 17-hydroxycorticosteroids (qv) or Porter–Silber chromogens.

See also: **cortisol**

POTASSIUM

Metabolism

This is the major cation of intracellular fluid. Along with sodium, it is absorbed throughout the small intestine. It is lost from the body in two ways:

(1) In intestinal secretions. Most of this potassium is however reabsorbed.

(2) In urine. Nearly all the potassium in the glomerular filtrate is reabsorbed in the proximal tubule. Potassium is resecreted in the distal tubule in exchange for sodium, a process which is stimulated by aldosterone.

The metabolism of hydrogen ions is closely linked with that of potassium ions, changes in the potassium concentration affecting the acid–base balance of the body. For instance in potassium depletion, H^+ enters the cells and an extracellular alkalosis develops.

Hypokalaemia

Potassium is involved in neuromuscular transmission and this is disturbed when the serum potassium is low. The results of this include cardiac arrhythmias and muscular weakness. Intracellular potassium deficiency leads to extracellular alkalosis, this in turn reducing the ionization of calcium and causing tetany. The main causes of hypokalaemia are:

(1) Reduced potassium intake, e.g. a result of starvation

(2) Loss of potassium by diarrhoea, vomiting or through intestinal fistulae.

(3) Loss of potassium in the urine, for instance by

  (a) Increased sodium–potassium exchange in the distal tubules. This occurs in primary and secondary aldosteronism and in Cushing's syndrome.

  (b) Reduced proximal tubular potassium reabsorption. This occurs in renal tubular disorders.

  (c) In patients on diuretics which inhibit proximal sodium reabsorption. More sodium is therefore made available for exchange with potassium in the distal tubule.

(4) Due to movement from the extracellular fluid into the cells, e.g. in diabetics treated with insulin or as a result of an alkalotic condition.

Hypokalaemia is treated by potassium administration, either orally or by intravenous infusion depending upon the severity.

Hyperkalaemia

The main danger of hyperkalaemia is that of cardiac arrest. Among the causes of hyperkalaemia are:

(1) Excessive potassium administration.

(2) Reduced renal excretion of potassium, e.g.

  (a) Renal failure.

  (b) Sodium depletion (less sodium available for exchange with potassium).

  (c) Addison's disease when insufficient aldosterone is secreted.

  (d) By aldosterone antagonists, such as spironolactone.

(3) Loss of intracellular potassium into the extracellular fluid in

conditions such as tissue damage or as a result of acidotic conditions.

Depending on the severity of the hyperkalaemia, different treatments are possible:

(1) Calcium infusion. This opposes the action of potassium on heart muscle.

(2) Glucose and insulin treatment, which increases the rate of potassium entry into cells.

(3) Use of oral ion-exchange resins which remove potassium from the body in exchange for sodium or calcium.

Measurement

In most laboratories, potassium is measured by atomic emission flame photometry (qv). However, in recent years ion-selective electrodes (qv) have been increasingly used.

Further reading: General list of clinical textbooks

PRE-ALBUMIN

A serum protein of molecular weight 61 000, which migrates in front of albumin on serum protein electrophoresis. It has similar functions to albumin but it is particularly important in the binding of thyroid hormones. Low serum levels are found in a variety of conditions including malignancies and liver diseases.

Further reading: Harris, R.I. and Kohn, J. (1974). The pre-albumin fraction. A useful parameter in the interpretation of routine protein electrophoresis. *J. Clin. Pathol.*, **27**, 986

PRECIPITIN REACTION

The reaction of antibody and antigen to form large insoluble complexes.

PRECISION

A measure of the reproducibility of analytical measurements. Precision can be expressed by variance, standard deviation, or coefficient of variation, the smaller these values the greater the precision.

Note that an analytical method can be precise without necessarily being accurate.

PREDNISOLONE TEST

A test which can be used to determine the cause of a high serum bilirubin level. A fall in the level of bilirubin after administration of prednisolone (a synthetic glucocorticoid) suggests hepatitis rather than obstruction as a cause of the jaundice. The reason for this is not clear.

PREGNANCY

A number of tests can be used for monitoring fetal well-being in pregnancy.

(1) Measurement of oestrogens in maternal urine. This is a measure of both fetal and placental function.

(2) Specific measurement of oestriol in plasma. This too is an index of feto-placental function.

(3) Specific measurement of progesterone in plasma or its urinary metabolite, pregnanediol. This measures placental function only.

(4) Measurement of human placental lactogen in plasma (a measure of placental function).

(5) Human chorionic gonadotrophin levels in blood and urine have been measured but are not considered to be a reliable guide to placental function, especially in late pregnancy, and are therefore infrequently used.

(6) Heat stable alkaline phosphatase. This is manufactured by the placenta and can be measured in blood as an index of placental function.

(7) Cystine aminopeptidase (oxytocinase) is an enzyme made by the placenta and it too can be measured in blood as an index of placental function.

(8) A number of placental proteins, e.g. pregnancy specific β_1-glycoprotein (SP_1), can also be measured in serum as an index of placental function.

(9) Amniotic fluid lecithin–sphingomyelin ratios can be measured to indicate the likelihood of respiratory distress syndrome.

See separate entries for each of these tests.

Further reading: Wilde, C.E. and Oakey, R.E. (1975). Scientific Review No. 3. Biochemical tests for the assessment of feto-placental function. *Ann. Clin. Biochem.*, **12**, 83

PREGNANCY-ASSOCIATED MACROGLOBULIN

An α-globulin, the level of which rises throughout pregnancy. It has been suggested that its level in the serum of patients with tumours can be correlated to the degree of malignant disease and can be used to monitor the progress of such conditions.

Further reading: Stimson, W.H. (1975). Variations in the level of a pregnancy-associated α-macroglobulin in patients with cancer. *J. Clin. Pathol.*, **28**, 868

Editorial (1975). Pregnancy-associated macroglobulin. *Lancet*, **2**, 1192

PREGNANCY-SPECIFIC β_1-GLYCOPROTEIN (SP_1)

One of a number of placental proteins which can be measured in serum as an index of placental function.

PREGNANEDIOL

This is the major urinary metabolite of progesterone. It can be measured in urine as an index of placental function in pregnant females, and to detect ovulation in non-pregnant females (see **progesterone**). It can be measured by gas chromatography, radioimmunoassay or chemical means (by its reaction with sulphuric acid following its isolation by column chromatography).

PREGNANETRIOL

A urinary metabolite of 17-hydroxyprogesterone, an intermediate in the biosynthesis of cortisol. Increased urinary excretion of pregnanetriol occurs in the 11-hydroxylase and 21-hydroxylase deficient forms of congenital adrenal hyperplasia, when plasma levels of 17-hydroxyprogesterone accumulate as a result of the metabolic block.

PREGNENOLONE

An intermediate in the biosynthesis of a number of steroid hormones including corticosteroids and androgens.

PRIMAQUINE SENSITIVITY

See: **glucose-6-phosphate dehydrogenase deficiency**

PRIMARY BILIARY CIRRHOSIS

An autoimmune disease in which there is diffuse intra-hepatic biliary obstruction. Mitochondrial autoantibodies can often be detected in the sera of affected patients. A raised serum IgM level can also be found.

PRIMARY STANDARD

See: **standards**

PRISMS

A means of isolating a narrow band of light wavelengths in some spectrophotometers.

PROBENECID

A drug which increases the renal excretion of uric acid and is therefore used in the treatment of gout.

PROGESTERONE

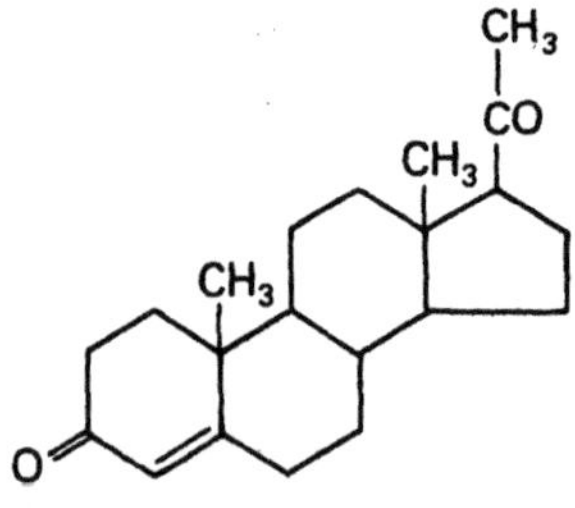

A steroid female sex-hormone secreted by the corpus luteum during the second half of the menstrual cycle when it prepares the uterus for the implantation of the embryo. It is also secreted by the placenta during pregnancy. The adrenals and testes also secrete small amounts of progesterone since it is an intermediate in the biosynthesis of corticosteroids and androgens. The major urinary metabolite of progesterone is pregnanediol.

Usefulness of progesterone estimations

Plasma progesterone levels can be measured to detect ovulation. If ovulation does not occur, there is no subsequent rise in the progesterone level. During pregnancy, progesterone levels in blood rise steadily and can therefore be used for monitoring progress.

Measurement

Progesterone can be measured by gas-chromatography, radioimmunoassay and competitive protein binding. A progesterone binding protein, obtained from pregnant guinea pigs, or transcortin, can be used for the competitive protein binding assays. Alternatively progesterone can be measured in urine as its metabolite, pregnanediol.

See also: **menstrual cycle**

PROLACTIN

A polypeptide hormone secreted by the anterior pituitary gland in both males and females. The highest serum levels occur in pregnancy, when, along with other hormones, it promotes lactation and mammary gland growth. Prolactin secretion, unlike other pituitary hormones, is limited by a hypothalamic prolactin-release inhibiting factor (PIF).

High serum levels in conditions other than pregnancy can cause galactorrhoea. Among the conditions associated with pathologically high serum prolactin levels are pituitary tumours and hypothalamic disorders. High levels can also occur in patients on drugs such as phenothiazines or methyldopa which reduce PIF secretion. Hypersecretion of prolactin can be treated with bromocriptine (qv).

Low serum prolactin levels are found in cases where there is pituitary dysfunction, e.g. Sheehan's syndrome (qv).

Prolactin is usually measured by radioimmunoassay.

Further reading: Editorial. (1977). Prolactin update. *Br. Med. J.*, **2**, 846

McNeilly, A.S. (1974). Prolactin and human reproduction. *Br. J. Hosp. Med.*, **12**, 57

PROLACTIN RELEASE INHIBITING FACTOR (PIF)

See: **prolactin**

PROLINE

An imino acid excreted in the urine in large amounts, along with hydroxyproline and glycine, in the inborn error, familial iminoglycinuria. This is due to abnormal renal tubular transport of these compounds.

Increased blood levels are found in the rare inborn errors, the hyperprolinaemias. The metabolic defect in type I hyperprolinaemia is an absence of proline oxidase, the first enzyme of the proline degradative pathway. A deficiency of Δ^1-pyrroline-5-carboxylic acid dehydrogenase, another enzyme involved in proline degradation, may be responsible for type II hyperprolinaemia.

Further reading: Scriver, C.R. (1978). Disorders of proline and hydroxyproline metabolism. In Stanbury, J.B., Wyngaarden, J.B. and Fredrickson, D.S. (eds.) *The Metabolic Basis of Inherited Disease.* 4th Edn., p.336. (New York: McGraw-Hill)

PROSTAGLANDINS

A group of compounds synthesized from fatty acids which affect smooth muscle, blood vessels and other tissues. It is possible that prostaglandins may be involved in the aetiology of Bartter's syndrome (qv).

Further reading: Russell, P.T., Eberle, A.J. and Cheng, H.C. (1975). Review: The prostaglandins in clinical medicine. A developing role for the clinical chemist. *Clin. Chem.*, **21**, 653

PROTEIN-BOUND IODINE (PBI)

See: **thyroxine**

PROTEIN-LOSING ENTEROPATHY

A rare syndrome in which there is an excessive loss of protein from the body into the gut, due to the intestinal wall being abnormally permeable to large molecules (in an analogous manner to nephrotic cyndrome, qv). It occurs in a number of gut conditions, e.g. when there is bowel ulceration. Hypoproteinaemia is therefore a feature of these conditions. It can be diagnosed by injecting radioactively labelled albumin, or a substance with a molecular weight similar to that of albumin into the blood and measuring its loss into the bowel. e.g. radioactively labelled polyvinylpyrrolidone.

PROTEINS

The proteins in plasma include enzymes, transport proteins, hormones, clotting factors and antibodies. Collectively they are responsible for the plasma oncotic pressure. Electrophoretically the proteins can be classified into a number of different fractions:

(1) Pre-albumin

(2) Albumin.

(3) α_1-Globulins

(4) α_2-Globulins

(5) β-Globulins

(6) γ-Globulins

(For more details see **albumin, globulin, pre-albumin**

Causes of a raised serum total protein level

(1) Dehydration.

(2) Paraproteinaemia.

(3) Chronic inflammatory diseases.

(4) Liver cirrhosis.

(5) Certain autoimmune diseases such as rheumatoid arthritis and systemic lupus erythematosus.

Causes of a decreased serum total protein level

(1) Overhydration.

(2) When there is decreased synthesis of protein, as in malnutrition, malabsorption and severe liver disease.

(3) When there is excessive loss of protein, as in nephrotic syndrome or protein-losing enteropathy.

Measurement of serum total protein

(1) Most serum protein measurements are based on the biuret reaction. This is the reaction between the peptide bonds of

proteins and cupric ions in alkaline solution to form a coloured chelation complex of unknown composition. It is called the biuret reaction because an analogous reaction takes place between cupric ions and the organic compound, biuret (NH_2–CO-NH-CO-NH_2). The intensity of the colour produced is proportional to the number of peptide bonds reacting.

(2) The Kjeldahl technique (qv) may be used as a reference method but it is too laborious for routine use.

(3) Refractometry (qv) can be used for a rapid estimation of the total serum protein.

Urinary proteins

Protein may be found in the urine in a number of conditions (see **proteinuria**). Urinary protein can be detected by assessing the turbidity after adding acid (e.g. sulphosalicylic acid) or by Albustix (qv).

Further reading: Freeman, T. (1970). Plasma proteins. *Br. J. Hosp. Med.*, **3**, 683

PROTEIN SELECTIVITY

See: **nephrotic syndrome**

PROTEINURIA

In a normal adult only a small amount of protein is excreted in the urine (up to 0.08 g/day). Orthostatic (postural) proteinuria, where proteinuria occurs in the upright position, can occur in apparently normal people, especially young adults. Pathological proteinuria has been classified into four types:

(1) *Glomerular proteinuria,* as in nephrotic syndrome (qv) where the renal glomeruli have an increased permeability to larger molecules.

(2) *Tubular proteinuria.* Normally the smaller molecular weight proteins, e.g. β_2-microglobulin, pass through the glomerulus and are reabsorbed by the renal tubules. Conditions such as pyelonephritis result in renal tubular damage and the increased excretion of lower molecular weight proteins in the urine.

(3) *Overflow proteinuria.* This occurs as a result of an accumulation of particular proteins in the blood, e.g. Bence-Jones protein in myeloma or myoglobin in crush syndrome (qv).

(4) *Nephrogenic proteinuria.* This includes renally derived proteins, e.g. Tamm–Horsfall protein or immunoglobulins, which can be secreted into the urine in increased amounts in nephritis.

Further reading: Hardwicke, J. (1979). Urinary proteins. In Milford Ward, A. and Whicher, J.T. (eds.) *Immunochemistry in Clinical Laboratory Medicine.* (Lancaster: MTP Press)

PROTHROMBIN

A plasma protein which is converted to thrombin, which then catalyses the conversion of fibrinogen to fibrin. The fibrin molecules polymerize to form the blood clot. Prothrombin can be measured in blood as the prothrombin time.

PROTOPORPHYRIA

See: **erythrohepatic protoporphyria**

PROTOPORPHYRIN

A porphyrin having methyl, vinyl and propionate side groups on the pyrrole rings. It is excreted in large amounts in faeces in erythrohepatic protoporphyria.

See also: **porphyrins**

PSEUDO-ADDISON'S DISEASE

A rare disease in which the renal tubules cannot respond to aldosterone.

PSEUDOCHOLINESTERASE

See: **cholinesterase**

PSEUDOGLOBULINS

The globulin fraction of serum proteins can be divided into euglobulins (globulins which precipitate between 28% and 33% ammonium sulphate concentration) and pseudoglobulins (globulins which precipitate between 33% and 50% ammonium sulphate concentration).

PSEUDOGOUT

A condition similar to gout, the major difference being that calcium pyrophosphate rather than uric acid is precipitated in the joints.

PSEUDOHYPOPARATHYROIDISM

A hereditary defect, the symptoms of which are similar to those of hypoparathyroidism. It is due to a failure of the hormone receptors in bone and kidney to respond to parathyroid hormone (PTH). It can be distinguished from hypoparathyroidism by the Ellsworth–Howard test (qv) which measures the phosphate excretion in urine in response to PTH. Cyclic AMP measurements (qv) can also be used to distinguish between these two conditions.

Further reading: Potts, J.T. Jr. (1978). Pseudohypoparathyroidism. In Stanbury, J.B. Wyngaarden, J.B. and Fredrickson, D.S. (eds.) *The Metabolic Basis of Inherited Disease*. 4th Edn., p. 1350. (New York: McGraw-Hill)

PYRIDOXINE

A water-soluble vitamin of the B group. As pyridoxal phosphate, it is a cofactor for amino acid decarboxylation and transamination reactions. A deficiency produces symptoms of skin roughening. The pyridoxine status of the body can be determined by the tryptophan loading test (qv).

PYROGEN TEST

A test of adrenocortical function based on the observation that bacterial endotoxins stimulate corticosteriod secretion. It is possible that this involves the stress control mechanism of cortisol secretion (see **corticotrophin releasing factor**). The test consists of injecting purified bacterial endotoxins and measuring the subsequent plasma cortisol levels. In normal subjects, an increase in the plasma cortisol level occurs. This response is diminished if there is dysfunction of the hypothalamic–pituitary–adrenal axis.

PYROGLOBULINS

Proteins which coagulate when serum is heated to 45–55 °C. They are found most often in myeloma but can also occur in other conditions such as chronic infections.

PYROPHOSPHATE

See: **calcium pyrophosphate**

PYRUVATE

An intermediate in the catabolism of glucose. If there is inadequate oxygen available, it is converted to lactate (see **lactate**). If adequate oxygen is available, it is decarboxylated to acetyl-CoA, a reaction which requires, thiamine pyrophosphate as a cofactor. Increased blood pyruvate levels are therefore found in cases of thiamine deficiency.

Measurement

(1) By its reaction with 2,4-dinitrophenylhydrazine to form the coloured dinitrophenylhydrazone derivative.

(2) Pyruvate can also be estimated enzymically by lactate dehydrogenase which converts pyruvate to lactate. At the same time NADH is oxidized to NAD and this can be followed spectrophotometrically.

PYRUVATE KINASE DEFICIENCY

An inborn error of metabolism in which there is a deficiency of pyruvate kinase; the glycolytic enzyme which converts phosphoenolpyruvate to ATP and pyruvate. This deficiency results in a reduced synthesis of ATP and a diminished capacity to cycle NAD in erythrocytes. This results in a haemolytic anaemia. The condition can be diagnosed by measuring the level of the enzyme in erythrocytes. It is inherited as an autosomal recessive.

Further reading: Valentine, W.N. and Tanaka, K.R. (1978). Pyruvate kinase and other enzyme deficiency hereditary haemolytic anaemias. In Stanbury, J.B., Wyngaarden, J.B. and Fredrickson, D.S. (eds.) *The Metabolic Basis of Inherited Disease.* 4th Edn., p. 1410. (New York: McGraw-Hill)

PYRUVATE METABOLISM TEST

A test, used mainly for the diagnosis of thiamine deficiency, which consists of giving an oral load of glucose and measuring the subsequent blood pyruvate levels (see pyruvate). In normal individuals only a small rise in the level occurs. A larger increase, however, occurs in thiamine deficiency and in diseases where there is disturbed carbohydrate metabolism such as liver diseases. An absent response may be found in untreated diabetics.

Q

QUALITY CONTROL

This ensures that reliable laboratory results are being produced. There are several ways of monitoring performance:

(1) Inter-laboratory quality control schemes in which different laboratories assay the constituents of a common sample.

(2) Intra-laboratory quality control schemes. These include:

 (a) Calculation of the daily mean for a given parameter. If the batches being analysed are sufficiently large, the daily mean should vary very little from day to day. The finding of a large change in the daily mean may indicate a fault in the analytical system.

 (b) Accuracy can be assessed by including a specimen with a known value in the batch.

 (c) Precision can be monitored by multiple analysis of a specimen which does not necessarily have a fixed value.

 (d) The cusum plot (qv) can be used to detect slight trends in the day to day quality of results.

Further reading: Whitby, L.G., Mitchell, F.L. and Moss, D.W. (1967). Quality control in routine clinical chemistry. In Bodansky, O. and Stewart, C.P. (eds.) *Advances in Clinical Chemistry*. Vol. 11, p. 66. (New York: Academic Press)

Whitehead, T.P. (1977). Advances in quality control. In Bodansky, O. and Latner, A.L. (eds.) *Advances in Clinical Chemistry*. Vol. 19, p. 175. (New York: Academic Press)

QUINALBARBITONE

An intermediate acting barbiturate.

See: **barbiturates**

R

RADIAL IMMUNODIFFUSION (MANCINI TECHNIQUE)

An immunochemical technique for the determination of specific proteins in biological fluids. It consists of placing the samples in wells cut out of a sheet of agar gel into which specific antiserum has been incorporated. Diffusion of the proteins from the well into the gel results in the formation of antibody–antigen complexes which precipitate in the form of a ring around the well. The concentration of the protein antigen is proportional to the area (and therefore the diameter squared) of the precipitin ring.

Further reading: Grant, G.H. and Butt, W.R. (1970). Immunochemical methods in clinical chemistry. In Bodansky, O. and Stewart, C.P. (eds.) *Advances in Clinical Chemistry*. Vol. 13, p. 383. (New York: Academic Press)

RADIOACTIVE IODINE

The three radioactive isotopes of iodine that are in common use are:

^{125}I: halflife 58 days, emits γ-radiation.
^{131}I: halflife 8 days, emits β- and γ-radiation.
^{132}I: halflife 2.3 hours emits β- and γ-radiation.

RADIOACTIVE IODINE NECK UPTAKE TESTS

An *in vivo* test for the diagnosis of thyroid disorders which consists of giving radioactive iodine to the patient and measuring its uptake by the thyroid gland, by means of a radioactive counter placed over the gland. Uptakes are usually measured after 4 or 24 hours. High uptakes occur in hyperthyroidism whereas low uptakes are found in hypothyroidism.

Two isotopes can be used:

(1) ^{131}I which has a halflife of 8 days and can be used for prolonged tests.

(2) ^{132}I which has a halflife of 2.3 hours and can therefore only

be used for 4 hour uptakes. However, it results in a much lower radiation dose being given to the patient.

If high uptakes are found, the test can be repeated after giving a dose of tri-iodothyronine (the T_3 suppression test, qv). This distinguishes hyperthyroidism from other possible causes of high uptake. Similarly if a low uptake has been found, a TSH stimulation test (qv) can help in the differential diagnosis of the hypothyroidism.

Further reading: Wellby, M.L. (1976). The laboratory diagnosis of thyroid disorders. In Bodansky, O. and Latner, A.L. (eds.) *Advances in Clinical Chemistry*. Vol. 18. p. 104. (New York: Academic Press)

RADIOACTIVITY MEASUREMENT

There are two commonly used techniques in clinical chemistry for the measurement of radioactivity.

Techniques based on the ionization of gases

In these techniques radiation causes the ionization of a gas contained in an enclosed chamber (an ionization chamber). A potential is applied across the gas by means of two electrodes. The ions produced from the ionization of the gas by the radiation are collected at the electrodes and the resultant current can be measured.

If a high voltage is applied, electrons formed by the primary ionization are accelerated to high speeds and cause secondary ionization of the gas in the chamber. The size of the pulse produced is proportional to the initial primary ionization. Such counters are therefore called proportional counters. They can be used for discriminating between different types of radiation (e.g. α-and β-radiation) which differ in the intensity of the primary ionization.

When the voltage is raised even further the secondary ionizations reach saturation and the size of the resulting electrical pulse is independent of the energy of the incident radiation. This is the Geiger–Müller region and such counters are known as Geiger–Müller counters. Both these and proportional counters can be used for measuring β-radiation. They are not suitable for γ-radiation because of the penetrating power of such radiation.

Scintillation counting

These techniques are based on the principle that ionizing radiation causes certain materials to emit flashes of light or scintillations. There are two different forms of scintillation counting:

(1) *Solid scintillation counting* This is when the radiation comes from a source external to the scintillator. Many scintillation counters use a solid crystal of sodium iodide, activated with thallium. When radiation strikes the crystal, electrons are excited and when they return to a more stable energy state, they emit energy as light. The light can be detected by a photomultiplier tube, and the number of pulses counted electronically. This type of counter is most suitable for γ-radiation.

(2) *Liquid scintillation counting* This can be used for counting low level β-radiation. The radioactive sample is dissolved in a vial with a primary scintillant, e.g. 2,5-diphenyloxazole (PPO), and a secondary scintillant. The latter enables light to be emitted of a wavelength to which the photomultiplier is most sensitive. Liquid scintillation is not without its problems. These include:

(a) Phosphorescence of the samples. This can be overcome by placing the vials in the dark for a certain length of time prior to counting.

(b) Quenching, i.e. a decreased number of scintillations caused by materials in the sample. Several methods of quench correction can be used.

(c) The photomultiplier tube itself may emit spurious pulses. This can be overcome by:

(i) Cooling the photomultiplier to reduce thermal emission.

(ii) Using a coincidence technique in which two photomultiplier tubes are used. A count is only accepted when the pulses from each tube are simultaneous.

Further reading: General list of analytical textbooks

RADIOALLERGOSORBENT TEST (RAST)

An immunochemical test, manufactured by Pharmacia, for the detection of specific IgE antibodies in allergic patients. A series of allergens (e.g. pollen) are each incorporated into separate paper discs and these are then mixed with the patients serum. Any specific IgE present binds to the appropriate paper disc. The amount of specific IgE binding to the paper can then be measured by adding radioactively labelled anti-IgE.

Further reading: Editorial. (1976). The RAST test. *Lancet,* **1,** 1061

RADIOIMMUNOASSAY (RIA)

Radioimmunoassays are techniques for the estimation of particular compounds, based on competition for binding to antibody between the compound in the sample and radioactively labelled compound. The more of the substance there is in the sample, the less of the radioactively labelled compound will bind to the antibody. Conversely the smaller the amount of the substance in the sample, the greater will be the amount of radioactively labelled compound that will bind to the antibody.

Radioimmunoassays can be used to measure a wide variety of hormones, drugs, proteins and other substances. The major advantages of radioimmunoassays are their high degree of specificity and sensitivity.

In order to set up a radioimmunoassay for a particular substance, two materials are required:

(1) *Specific antiserum to the substance being measured.* Many substances, such as hormones, are however too small to stimulate antibody production in the animal and therefore have to be conjugated to an inert protein in order to render them immunogenic.

(2) *Radioactively labelled compound.* ^{3}H, ^{14}C, ^{125}I and ^{131}I have been used for labelling. The iodine isotopes are used particularly for the labelling of proteins and peptides. Iodination can be achieved by oxidizing iodide (I^-) to iodine (I_2) with chloramine-T. The I_2 then reacts with tyrosine residues in the protein or peptide. A number of other techniques are also available for iodination.

In the radioimmunoassay procedure itself, the patient's sample or standard is incubated with the labelled material and the specific antiserum. At the end of the incubation period, the reaction mixture contains both free and bound labelled material and these have somehow to be separated. This can be achieved in one of several ways:

(1) Removal of the free antigen by adsorption using such materials as charcoal, ion-exchange resin or Fuller's earth.

(2) Precipitation of the antigen bound to antibody by protein precipitants such as ethanol or ammonium sulphate.

(3) Precipitation of the antigen bound to antibody by using a second antibody. If, for example, the primary antibody is obtained from sheep, the second antibody can be antiserum raised against sheep immunoglobulins using a different animal such as a horse.

(4) Another approach is to use solid phase antibodies. The antibodies may be attached to the inside of a plastic tube. At the end of the incubation period, the assay mixture is removed from the tube and the tube itself is counted. Antibodies may also be covalently bound to a solid material, such as cellulose or glass beads, in which case they can be removed by centrifugation, or attached to magnetic particles, in which case they can be removed magnetically.

Having separated the free and the bound antigen, one or the other is counted. A calibration graph (e.g. % of antigen bound vs concentration of standards) is prepared from the standards. As this may not be linear in shape, various mathematical treatments are possible in order to transform it into a straight line graph. This enables more automated approaches to the calculation of results.

Further reading: Skelley, D.S., Brown, L.P. and Besch, P.K. (1973). Radioimmunoassay. *Clin. Chem.*, **19**, 146

Radioimmunoassay and saturation analysis. (1974). *Br. Med. Bull.*, **30**, 1

RAST

See: **radioallergosorbent test**

RECEPTOR BINDING ASSAYS

A technique analogous to radioimmunoassay or competitive protein binding, the difference being that labelled and unlabelled material compete with each other for binding to a specific tissue receptor (e.g. the oestradiol receptor in uterine tissue) rather than an antibody or serum binding protein.

Further reading: Rees Smith, B. (1977). Membrane receptors for polypeptide hormones. In Bodansky, O. and Latner, A.L. (eds.) *Advances in Clinical Chemistry*. Vol. 19, p. 91 (New York: Academic Press)

REDUCING SUBSTANCES IN URINE

Among the urinary substances which will give a positive Benedict's test for reducing substances are:

glucose
glucuronates (metabolites of drugs)
lactose
fructose
galactose
pentoses
salicyluric acid (metabolite of salicylate)
homogentisic acid
urates and creatinine, when present in high concentrations

REFERENCE METHOD

A method which gives the best possible analytical result for a particular substance, without necessarily being suitable for routine use in a laboratory. It provides a yardstick with which more routine methods may be compared.

REFERENCE RANGE

A term which for most purposes can be thought of as the normal range. The term is preferred to normal range because the latter is usually represented by the mean ± 2SD (i.e. 95%) of a normal population and therefore, by definition, 5% of normal individuals will be outside the "normal" range. The expression 'reference range' therefore avoids the implication that any value in the range is normal and outside the range is abnormal. Furthermore the samples that are used in constructing many "normal" ranges do not come from normal people but instead are often drawn from hospital patients or patients attending clinics.

See also: **normal range**

REFRACTOMETRY

A technique in which the refraction of light is measured. The refractive index of a solution is related to the number and type of dissolved solutes in it (as is specific gravity). Measurements on urine can therefore be made as an alternative to specific gravity determination.

Serum refractive index measurements can also be made in order to determine serum protein concentrations. These are based on the assumption that the concentrations of small molecular weight substances do not vary signficantly from specimen to specimen and that the differences in the refractive index are due to differences in the protein concentration.

REFSUM'S DISEASE (PHYTANIC ACID STORAGE DISEASE)

A hereditary neurodegenerative disorder of lipid metabolism in which the branched-chain fatty acid, phytanic acid, accumulates in the tissues, particularly the liver and kidney. It is due to a deficiency of phytanic acid α-hydroxylase, an enzyme involved in the oxidation of phytanic acid. The phytanic acid originates from dietary phytanic acid and also from dietary phytol, a component of chlorophyll. Treatment therefore consists of giving diets low in phytol and phytanic acid.

Further reading: Steinberg, D. (1978). Phytanic acid storage disease: Refsum's syndrome. In Stanbury, J.B., Wyngaarden, J.B. and Fredrickson, D.S. (eds.) *The Metabolic Basis of Inherited Disease*. 4th Edn., p. 688. (New York: McGraw-Hill)

REGAN ISOENZYME

A heat stable alkaline phosphatase isoenzyme similar to placental isoenzyme. It is found in some patients with carcinoma. Regan isoenzyme is named after the first patient in whom it was found.

REINSCH TEST

A test for the detection of arsenic and other metallic elements. It is based on the reduction of arsenic by metallic copper in the presence of acid. The arsenic is deposited on the copper as a dark film. The test can also be used for the detection of antimony, bismuth, mercury and selenium since the oxidized forms of these elements can also be reduced by copper.

RENAL CONCENTRATION TEST

See: **water deprivation test**

RENAL FAILURE

Renal failure can be classified as either acute or chronic.

Acute renal failure

This can occur as a result of a number of factors. These include acute glomerulonephritis, haemorrhages, septicaemia, mismatched transfusions or post renal obstruction (e.g. stones).

There are usually two distinct phases to the condition:

(1) In the first phase there is reduced urine output (oliguria). Plasma urea and potassium levels increase and an acidosis results as a consequence of the failure to excrete hydrogen ions.

(2) The oliguric phase is followed by a diuretic (recovery) phase in which there is a high output of dilute urine. This is mostly glomerular filtrate as the tubules are not yet working. Water, sodium and potassium are lost from the body during this state.

Chronic renal failure

Among the most important causes of chronic renal failure are chronic glomerulonephritis, chronic pyelonephritis and hypertension. Various tubular disorders, such as heavy metal poisoning, Wilson's disease and tubular damage caused by hypercalcaemia or hyperuricaemia, may also lead to generalized chronic renal failure.

The biochemical features of chronic renal failure are those of:

(1) Urea, creatinine, urate and phosphate retention

(2) Failure to eleminate acids, leading to acidosis.

(3) Hyperkalaemia.

(4) Hypocalcaemia, possibly due to the failure of the kidney to hydroxylate vitamin D.

(5) Anaemia, due to failure to produce erythropoietin.

Further reading: General list of clinical textbooks

RENAL FUNCTION TESTS

These will be considered only briefly. For a more detailed account of the compound or test see under the appropriate headings.

(1) Blood urea or creatinine.

(2) Specific gravity or osmolality measurements on urine.

(3) Measurement of glomerular filtration rate by determination of the clearance of creatinine, inulin, mannitol or [^{51}Cr]EDTA.

(4) Measurement of the renal concentrating ability by the water deprivation test.

(5) Measurement of the kidneys' ability to excrete a water load by the water excretion test.

(6) Measurement of the ability of the kidneys to secrete an acid load by the ammonium chloride loading test.

(7) Measurement of the renal plasma flow by measuring the clearance of low levels of *p*-aminohippuric acid.

(8) Measurement of the tubular secretory capacity by phenolsulphonephthalein excretion.

(9) Measurement of the ability of the tubules to respond to parathyroid hormone by the Ellsworth–Howard test.

Further reading: Mitchell, F.L., Veall, N. and Watts, R.W.E. (1972). Scientific Review No. 2. Renal function tests suitable for clinical practice. *Ann. Clin. Biochem.*, **9**, 1

RENAL OSTEODYSTROPHY

A disease of calcium metabolism in which there are changes in bone structure associated with chronic renal failure. A factor in the aetiology of the condition may be the failure to hydroxylate vitamin D by the kidney. The disease may be treated by giving large doses of vitamin D.

RENAL PLASMA FLOW

Renal function depends on the renal blood flow. The determination of the clearance of a substance which is removed from the body by both filtration at the glomerulus and secretion by the tubules, therefore, gives an indication of the renal plasma flow. *p*-Aminohippuric acid is such a substance. At low plasma levels it is almost completely removed from the plasma in a single passage through the kidneys. Diodrast is another substance which can be used to measure renal plasma flow.

Further reading: Mitchell, F.L., Veall, N. and Watts, R.W.E. (1972). Scientific Review No. 2. Renal function tests suitable for clinical practice. *Ann. Clin. Biochem.*, **9**, 1

RENAL TUBULAR ACIDOSIS

A condition, which may be congenital or acquired, in which the renal tubules fail to acidify the urine normally. It can be diagnosed by the ammonium chloride loading test, which consists of giving an oral load of ammonium chloride and measuring the acid output in the urine. Failure to produce an acid urine is suggestive of renal tubular acidosis.

See also: **ammonium chloride loading test**

Further reading: Seldin, D.W. and Wilson, J.D. (1978). Renal tubular acidosis. In Stanbury J.B., Wyngaarden, J.B. and Fredrickson, D.S. (eds.) *The Metabolic Basis of Inherited Disease*. 4th Edn., p. 1618. (New York: McGraw-Hill)

RENIN-ANGIOTENSIN SYSTEM

Renin is a proteolytic enzyme secreted by the renal juxtaglomerular apparatus in response to reduced renal blood flow. In the blood, it acts on the plasma protein angiotensinogen to form the decapeptide angiotensin I. Angiotensin I is further split to angiotensin II by a peptidase located mainly in the lungs. Angiotensin II has two actions, one of which is to cause vasoconstriction, the other being the stimulation of aldosterone secretion, an action which results in sodium and subsequently, water retention. These two actions therefore can compensate for the original reduced renal blood flow.

Renin assays can be useful in distinguishing between primary and secondary aldosteronism. It can be measured by its enzymic action on angiotensinogen, the product, angiotensin I, being measured by radioimmunoassay.

Further reading: Editorial (1975). When to measure renin. *Lancet*, **1**, 783

RESIN UPTAKE TEST

See: **T_3 uptake test**

RETICULIN ANTIBODIES

Autoantibodies found in many patients with coeliac disease.

RETINOL

See: **vitamin A**

RETINOL-BINDING PROTEIN

The protein which carries retinol (vitamin A) in the blood. It has a molecular weight of 21 000 and in serum it is associated with the prealbumin fraction. Measurement of the protein is reported to be useful in assessing vitamin A deficiency and malnutrition.

REVERSE PHASE CHROMATOGRAPHY

In partition chromatography (qv), the stationary phase is usually a polar solvent (e.g. water), while the mobile phase is usually an organic solvent. Thus non-polar solutes migrate faster than polar solutes. In reverse phase chromatography, however, it is the stationary phase which is organic and the mobile phase which is aqueous. This is achieved by coating the support material (paper or thin layer plate) with a non-polar material such as paraffin or vaseline. Reverse phase chromatography enables a better separation of lipid soluble compounds than conventional partition chromatography.

REVERSE T_3

The deiodination of thyroxine in peripheral tissues results in the formation of 3,3′,5-tri-iodothyronine (T_3) or 3,3′5′-tri-iodothyronine (reverse T_3). Unlike T_3, which is more physiologically active than thyroxine itself, reverse T_3 has no physiological activity. It is possible that the ratio of T_3 to reverse T_3 may vary in various physiological and pathological conditions. For example, reverse T_3 is found in high concentrations in amniotic fluid. After birth, the serum reverse T_3 level falls while the level of normal T_3 increases. It is therefore possible that reverse T_3 may be a marker of congenital hypothyroidism in fetuses. Specific radioimmunoassays are available for the determination of reverse T_3.

RHEUMATOID ARTHRITIS

An immune complex disease in which the joints are badly affected. In many patients, rheumatoid factor can be demonstrated in the serum.

RHEUMATOID FACTOR

An autoantibody belonging to the IgM class, found in the serum of many patients with rheumatoid arthritis. It reacts against IgG. Several agglutination reactions can be used for its detection.

RIBOFLAVIN

A member of the water-soluble B complex of vitamins. It is a constituent of the flavin coenzymes, flavin mononucleotide (FMN) and flavin adenine dinucleotide (FAD), which participate in biological reduction–oxidation reactions. Riboflavin is usually obtained in the diet from plant sources. Deficiency of the vitamin (ariboflavinosis) can result in a rough scaly skin and oral, anal and vaginal lesions.

Measurement

Riboflavin can be measured by microbiological or fluorimetric methods. The erythrocyte enzyme, glutathione reductase, requires FAD as a cofactor and its measurement in the presence and absence of added FAD is therefore considered a good indicator of riboflavin status.

See also: **glutathione reductase**

RICKETS

A deficiency of vitamin D in children causes the condition known as rickets in which there is softening and irregular growth of bones. The disease can be treated with vitamin D. Occasionally, however, vitamin D resistant rickets may be encountered. This is primarily a failure of the renal tubules to reabsorb phosphate.

ROCKET ELECTROPHORESIS

See: **crossed electrophoresis**

ROTHERA'S TEST

A test for the detection of ketone bodies in urine based on their reaction with sodium nitroprusside to produce a purple colour.

S

SALICYLATE

Aspirin (acetylsalicylic acid) is a commonly used analgesic drug which is also used extensively for treating rheumatic and other disorders. When ingested, it is rapidly hydrolysed to salicylic acid and hence is commonly referred to as salicylate.

Toxicology

Aspirin is encountered frequently in the laboratory in overdose cases. It directly stimulates the respiratory centre in the brain causing hyperventilation and a resultant respiratory alkalosis. Later, a metabolic acidosis is superimposed on this and the acid–base picture becomes more complex. Potassium levels should therefore be monitored closely. Removal of the drug from the circulation is encouraged by forced alkaline diuresis.

Measurement

Salicylate is most commonly determined by the Trinder method which is based on the formation of a violet coloured complex between ferric ion and the phenolic group of salicylic acid.

Further reading: General list of analytical and clinical text-books

SALICYLURIC ACID

A metabolite of salicylate, being a conjugate of this compound and glycine. It can be detected as a reducing substance in urine, following ingestion of salicylates

SALIVA ELECTROLYTES

In cystic fibrosis, the concentration of sodium and chloride in saliva may be raised in a similar manner to the sweat electrolytes. However, this is an inconsistent finding and the results are not considered as reliable as sweat electrolytes.

SANFILIPPO SYNDROME

A mucopolysaccharidosis (qv).

SARCOIDOSIS

A condition in which hypercalcaemia can occur, possibly due to vitamin D sensitivity.

SATURATION ANALYSIS

A class of analytical techniques in which the substance to be measured reacts with a specific binding reagent (e.g. immunoglobulin or specific binding protein) of limited capacity, i.e. the binding reagent is saturated. Thus two fractions of that substance exist, a bound and a free fraction. If the substance to be measured competes with radioactively labelled substance for binding, measuring the radioactivity in the free or bound fraction and relating it to a set of standards allow the amount of the substance originally present in the sample to be ascertained. Competitive protein binding and radioimmunoassay are therefore examples of saturation analysis. Other labelled material can be used as an alternative to radioactive labels, e.g. enzymes in enzyme-immunossay.

See also: **competitive protein binding, enzyme-immunoassay, radioimmunoassay**

SCHEIE SYNDROME

A mucopolysaccharidosis (qv).

SCHILLING TEST

A test that can be used for the diagnosis of vitamin B_{12} malabsorption. A large dose of unlabelled vitamin B_{12} is first given to the patient to saturate the body stores and to ensure that the subsequent radioactive dose is excreted in the urine. Radioactive ^{57}Co-labelled vitamin B_{12} is then given orally and its urinary excretion is measured. If there is deficient absorption, the test can be repeated by giving an oral dose of the labelled vitamin together with intrinsic factor. If absorption is now normal, it suggests that the patient is failing to produce adequate amounts of intrinsic factor. However, if there is still deficient absorption, it suggests a generalized malabsorption due to intestinal disease. Both these tests may be performed simultaneously using two isotopic forms of vitamin B_{12} (^{57}Co- and ^{58}Co-labelled) one free and one bound to intrinsic factor.

See also: **vitamin B_{12}**

SCHLESINGER'S TEST

A test that can be used for the detection of urobilinogen or urobilin in urine or faeces. First iodine is added and this oxidizes urobilinogen to urobilin. Zinc acetate is then added and the fluorescence of the zinc urobilin, formed as a result, is assessed.

SCHUMM'S TEST

A test for methaemalbumin in plasma. It consists of covering the sample with ether and then adding ammonium sulphide solution. If methaemalbumin is present, an albumin haemochromogen is formed and this can be detected spectroscopically by its characteristic absorption band at 558 nm.

SCINTILLATION COUNTING

See: **radioactivity measurement**

SCRIVER TEST

A screening technique which consists of the paper chromatography of serum amino acids. It is used for the detection of inborn disorders of amino acid metabolism.

SECONDARY STANDARD

See: **standards**

SECRETIN

A polypeptide hormone, secreted by the duodenum, which stimulates the pancreatic production of fluid rich in bicarbonate. Gastric acid stimulates secretin release.

SECRETIN-CHOLECYSTOKININ-PANCREOZYMIN STIMULATION TEST

An *in vivo* test for the assessment of pancreatic function which consists of giving secretin (which stimulates pancreatic bicarbonate production) and cholecystokinin–pancreozymin (which stimulates the secretion of pancreatic enzymes). Pancreatic fluid is collected and analysed for bicarbonate and one of the pancreatic enzymes (trypsin, amylase or lipase). Low outputs occur in pancreatic disease.

Further reading: Gowenlock, A.H., (1977). Scientific Review No. 4: Tests of exocrine pancreatic function. *Ann. Clin. Biochem.*, **14**, 61

SECRETIN STIMULATION TEST

A test for the assessment of pancreatic function, which consists of giving secretin to the patient and then measuring the bicarbonate content of the pancreatic fluid. Low outputs occur in pancreatic disease.

Further reading: Gowenlock, A.H. (1977). Scientific Review No. 4. Tests of exocrine pancreatic function. *Ann. Clin. Biochem.*, **14**, 61

SECRETORY IgA

The IgA in secretory fluids is present in the form of a dimer, known as secretory IgA, in which the two immunoglobulin molecules are linked to a small protein (the secretory piece).

SECRETORY PIECE

See: **secretory IgA**

SELIWANOFF'S TEST

A test for the detection of fructose in urine, based on its reaction with resorcinol in hydrochloric acid to give a red colour.

SEMINAL FLUID FRUCTOSE

Over 50% of the seminal fluid is thought to originate from seminal vesicle secretion. Fructose is found in high concentrations in seminal vesicle secretions. Therefore, measurement of this sugar in semen enables the contribution of seminal vesicle secretion to the total composition of the semen to be determined. The activity of the seminal vesicles may be related to the pathology of the male reproductive tract and this may be responsible for certain cases of male infertility.

SEMINOMA

A malignant tumour of the testis. Among the biochemical findings in this condition is increased pituitary gonadotrophin secretion.

SEQUESTRENE

See: **ethylenediamine tetracetic acid**

SEROMUCOIDS

See: **mucoproteins**

SEROTONIN

See: **5-hydroxytryptamine**

SERUM

The fluid remaining after blood has clotted and the clot has been removed. This contrasts with plasma which is obtained when blood clotting has been prevented. Plasma, unlike serum, therefore contains fibrinogen.

SEVERINGHAUS ELECTRODE

A means of measuring PCO_2

See: **carbon dioxide**

SEX HORMONE BINDING GLOBULIN (SHBG)

A plasma protein which binds testosterone and oestrogens. In hyperthyroidism, the increased thyroid hormone levels induce the synthesis of SHBG. SHBG binds testosterone more avidly than oestrogens and the proportional increase in the free oestrogen levels in hyperthyroid males can cause gynaecomastia. Estimations of SHBG may also be useful in hirsute women whose plasma testosterone levels are normal. In such cases the SHBG may be found to be low and this results in greater proportions of the free (metabolically active) testosterone which accounts for the symptoms.

SEX-LINKED AGAMMAGLOBULINAEMIA (BRUTON'S DISEASE)

A genetic condition in which there is an almost complete absence of serum immunoglobulins. The defect is carried on the X chromosome and therefore it is males that are affected. Patients can be treated by γ-globulin therapy in order to prevent repeated infections.

See also: **agammaglobulinaemia**

SHEEHAN'S SYNDROME

A cause of panhypopituitarism due to pituitary infarction occuring at parturition.

See also: **hypopituitarism**

SIA TEST

A test for the detection of increased levels of IgM in serum. It consists of adding a drop of the serum to water or a dilute salt solution. A positive reaction occurs if a white precipitate is formed which re-dissolves on addition of sodium chloride.

SICKLE CELL ANAEMIA

A hereditary disease, usually found in negroes, in which there is an abnormal haemoglobin present in red cells which aggregates and precipitates out of solution. This causes the cells to assume the shape of a sickle. The disease presents as a chronic haemolytic anaemia. The abnormal haemoglobin is HbS which differs from normal haemoglobin in having one glutamic acid residue replaced by valine in the β chain. Sickle cell disease is the commonest haemoglobinopathy (qv).

SIDEROPHILIN

See: **transferrin**

SIMMOND'S DISEASE

Adult panhypopituitarism caused by such conditions as pituitary infiltration by neoplasms, trauma or pituitary infarction.

See also: **hypopituitarism**

SI UNITS

In the International System of Units (SI), there is a basic unit for each physical quantity. The basic unit of length is the metre, the basic unit of time is the second, the basic unit of mass is the kilogram and the basic unit for the amount of a substance is the mole. Other units can be derived from these.

SJØGREN'S SYNDROME

An immune-complex disease, usually occuring in middle-aged women, in which there are the symptoms of rheumatoid arthritis together with oral, ocular, and other lesions. Rheumatoid factor can be detected in the serum of the majority of patients.

SKEW DISTRIBUTION

An assymetrical distribution curve where the mean is different from the median.

SMOOTH MUSCLE ANTIBODIES

Autoantibodies which react with a component, possible actomyosin of smooth muscle cells. They can be found in the sera of patients with infectious mononucleosis, chronic active hepatitis, primary biliary cirrhosis and certain malignant conditions.

SODIUM

The major cation of extracellular fluid, where it makes the biggest contribution towards osmotic pressure.

Control of sodium metabolism

Sodium intake in the intestine is probably not under any active control. Aldosterone is probably the most important factor in the control of sodium excretion. This adrenal steroid hormone is secreted in response to the renin–angiotensin system (qv) and it promotes sodium reabsorption in the renal tubules (among other places) in exchange for potassium or hydrogen ions.

Disturbances of sodium metabolism

Primary disturbances of plasma sodium cause a change in the plasma osmolality and this results in a redistribution of the body water. The opposite is true in that if there is a primary disturbance of water metabolism, there is a subsequent disturbance of sodium metabolism. Thus water and sodium metabolism are closely linked. Nevertheless it is possible to identify certain conditions in which the primary disorder is one of sodium metabolism.

(1) *Primary sodium deficiency.* This is found in Addison's disease when there is inadequate secretion of aldosterone. Sodium deficiency can also occur when the fluid lost by vomitting, sweating, diarrhoea or through a fistula, is replaced by fluids low in sodium.
Hyponatraemia can occur in these conditions. Note that hyponatraemia can also occur in conditions where the primary disorder is of water metabolism (see **water-balance**), e.g. inappropriate ADH secretion.

(2) *Primary sodium excess.* This occurs in primary aldosteronism (Conn's syndrome) when there is an inappropriate secretion of aldosterone. Hypokalaemia, possibly accompanied by hypernatraemia, are features of this condition. Secondary aldosteronism, when aldosterone is secreted in conditions where there is stimulation of the renin–angiotensin system by reduced renal blood flow (e.g. hypoproteinaemic states or cardiac failure), can also be considered as a cause of abnormal sodium metabolism. Hypernatraemia, however, is not a feature of this condition and the plasma sodium level may even be low.

It should be noted that hypernatraemia can occur in conditions in which the primary disorder is that of water depletion, e.g. loss of fluid (in urine as occurs in diabetes insipidus) or deficient water intake, (see **water balance**).

Measurement

Sodium is usually measured by atomic emission flame photometry. However, in some of the latest analytical instruments, it is measured by ion-selective electrodes.

Further reading: Editorial (1976). Hyponatraemia. *Lancet,* **1**, 1334
Editorial (1978). Hyponatraemia. *Lancet,* **1**, 642
Thompson, F.D. (1979). Hyponatraemia. *Br. J. Hosp. Med.,* **21**, 46

SOFT-TISSUE PLASMACYTOMA

A condition which, in its later stages, can sometimes be indistinguishable from myeloma, but differs from it in that the primary monoclonal proliferation of plasma cells occurs in sites other than the bone marrow (usually the upper airway passages). A paraprotein can often be found in such cases.

Further reading: Wiltshaw, E. (1971). Extramedullary plasmacytoma. *Br. Med. J.,* **3**, 327

SOMATOMEDIN

A small peptide, synthesized in the liver, which mediates the action of growth hormone on skeletal growth.

SOMATOSTATIN (GROWTH HORMONE RELEASE INHIBITORY HORMONE)

A polypeptide hormone, originally isolated from the hypothalamus, which inhibits growth hormone secretion.

See also: **human growth hormone**

SOMATOTROPHIN

See: **human growth hormone**

SOMOGYI UNIT

A unit in which amylase activity may be expressed.

SORBITOL DEHYDROGENASE

An enzyme which catalyses the reaction:

$$\text{sorbitol} + \text{NAD} \rightleftharpoons \text{fructose} + \text{NADH}$$

Large amounts of this enzyme are found in the liver and therefore increased serum levels are usually indicative of hepatic disease. It can be measured spectrophotometrically by following the increase in NADH production

Further reading: Wilkinson, J.H. (1976). *The Principles and Practice of Diagnostic Enzymology*. (London: Edward Arnold).

SPECIFIC GRAVITY

Urine specific gravity measurements can be used to assess the degree of urine concentration, as, for instance, in the water excretion or water dilution tests (qv). Protein and glucose in the urine, however, increase the specific gravity. In recent years, osmometry has replaced specific gravity measurements.

Plasma specific gravity can be measured as an indication of the total protein concentration (see **proteins**).

SPECTROPHOTOMETRY

A widely used technique which consists of measuring the amount of light absorbed by molecules in solution, using a spectrophotometer. This instrument consists of the following components.

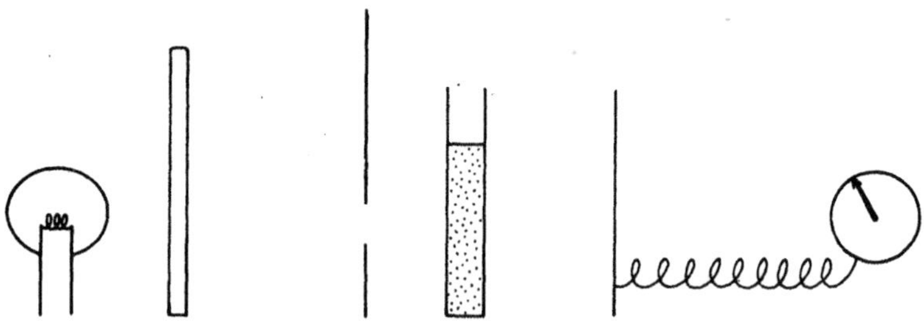

(1) A light source which is usually a tungsten lamp for visible light, a hydrogen or deuterium lamp for ultraviolet light or a quartz-iodide lamp for both regions.

(2) A monochromator which isolates a narrow band of light. Glass or gelatin filters, interference filters, diffraction gratings or prisms can be used to isolate a narrow wavelength band of light.

(3) A cuvette into which the solution is placed. It can be made from glass, quartz (silica) or plastic. Glass cuvettes are not suitable for work in the ultraviolet region. A proportion of the light falling on the solution is absorbed and the unabsorbed (transmitted) light passing through the solution is detected by:

(4) A photodetector such as a barrier layer cell or a photomultiplier tube.

(5) A meter, digital display, chart recording or some other means by which the electric current produced by the photodetector is displayed.

See also: **atomic absorption** for an account of atomic absorption flame spectrophotometry.

Further reading: Stevens, J. *et al.* (1975). Atomic and molecular absorption theory and practice as applied to clinical chemistry. *Med. Lab. Technol.*, **32**, 183

SPECTROSCOPY

A technique for the detection and identification of substances by examination of their characteristic absorption spectra.

SPHINGOMYELINS

These are lipids containing a fatty acid, phosphoric acid, choline and the amino alcohol, sphingosine. They are deposited in the tissues in Niemann–Pick disease, a disorder of lipid storage. In some forms of this disease, a deficiency of the lysosomal enzyme, sphingomyelinase, can be demonstrated.

Sphingomyelins are encountered in another aspect of clinical chemistry, that of lecithin–sphingomyelin ratios (qv).

SPIN IMMUNOASSAY

A technique which is similar to radioimmunoassay, the difference being that the antigen is labelled with a free radical such as nitroxide. When the antibody combines with the labelled antigen, the free radical is immobilized and broad spectral peaks are observed on an electron spin resonance spectrometer. When the labelled antigen is displaced by unlabelled antigen from the test sample, a sharp peak is produced. In this way, the amount of antigen in the patient's sample can be measured.

SPIRONOLACTONE

A drug which antagonizes the action of aldosterone in the renal tubules and which therefore can be used as a diuretic. Thus potassium is retained by the body and hyperkalaemia may develop as an undesirable side-effect.

SPRUE

A disease of intestinal malabsorption with steatorrhoea, associated with the tropics.

STANDARD BICARBONATE

A means of expressing the plasma bicarbonate level. It is the concentration of bicarbonate in plasma when fully oxygenated blood has been equilibrated at 37 °C and at a PCO_2 of 5.33 kPa (40 mmHg).

The plasma bicarbonate level is normally a reflection of both the erythrocyte buffering mechanisms and the renal acid–base homeostatic mechanisms. The former affects the actual bicarbonate level but not the standard bicarbonate. The standard bicarbonate gives a measure of the non-respiratory contribution and its measurement is therefore useful in acute respiratory disorders when a metabolic component is involved.

For determination see **Astrup**

STANDARD DEVIATION (SD)

This is a measure of the spread of a series of observations. The shape of a normal (Gaussian, q.v.) distribution curve is defined by its mean and standard deviation.

$$SD = \sqrt{\frac{\Sigma (x - \bar{x})^2}{n - 1}}$$

where n = the number of observations, x = the value of each observation, and $\bar{x}$ = the mean of the observations.

68.3% of the observations fall within ± 1SD
95.4% of the observations fall within ± 2SD
99.7% of the observations fall within ± 3SD

STANDARD ERROR OF THE MEAN

An estimate of how closely the mean of a sample is to the mean of the population from which the sample is taken. If a series of samples are taken from a population, the means of the samples has a normal distribution, and, from this, a standard deviation can be calculated. This particular standard deviation is smaller than the standard deviation of the population as a whole and is called the standard error of the mean.

It can be calculated from the formula

$$\text{standard error} = \frac{\text{standard deviation of the population}}{\sqrt{n}}$$

where n is the size of the sample.

Therefore the greater the value of n, the smaller the standard error, i.e. the closer the sample mean is to the mean of the population.

STANDARDS

Primary standard

A substance prepared or analysed by a method of high accuracy and precision, enabling it to be used as a reference for other standards.

Secondary standard

A material (e.g. serum), whose composition has been determined in relation to a primary standard and which may be used to prepare a calibration graph.

STARCH GEL ELECTROPHORESIS

A form of electrophoresis in which molecules are separated on the basis of both their charge and their molecular size.

STEIN-LEVENTHAL (POLYCYSTIC OVARY) SYNDROME

A syndrome associated with fibrocystic ovaries in which there is deficient conversion of androgens to oestrogens. The symptoms therefore include infertility, secondary amenorrhoea and hirsutism. In some, but not all, cases the plasma testosterone and androstenedione levels are raised and there may be an increased excretion of 17-oxosteroids.

STERCOBILIN (UROBILIN)

The compound formed by the oxidation of urobilinogen by colonic bacteria.

STERCOBILINOGEN

One of the compounds formed by the reducing action of intestinal bacteria on bilirubin. Together with the other bilirubin reduction products, mesobilirubinogen and urobilinogen, it forms the group of compounds known collectively as faecal urobilinogen. Stercobilinogen undergoes subsequent oxidation to stercobilin (urobilin).

STEROID SUPPRESSION TEST

A test that can be used for the differential diagnosis of hypercalcaemia. Large doses of cortisone or hydrocortisone normally cause elevated serum calcium levels to fall to within normal limits in nearly all cases of hypercalcaemia except primary or tertiary hyperparathyroidism. The reason for this is not known.

STIMULATION TESTS

These are a class of *in vivo* tests which are used mainly for diagnosing deficient hormone secretion by a gland. They consist of administering a trophic hormone or a substance that acts as a trophic hormone and then measuring hormone production by the target gland. A failure of the hormone level to rise indicates a primary abnormality of that gland.

The synacthen stimulation test illustrates the principle behind these tests. In normal persons, administration of synacthen (a synthetic form of ACTH) results in stimulation of the adrenal cortex and a subsequent rise in the plasma cortisol level. When there is adrenocortical hypofunction, as in Addison's disease, the rise in the plasma cortisol level does not occur or is markedly reduced.

See also: **suppression tests**

STREET-CLOSE UNIT

A unit in which amylase activity can be expressed.

STRISOWER CLASSIFICATION

A system for classifying the hyperlipoproteinaemias based on the behaviour of the lipoproteins in the ultracentrifuge.

See also: **hyperlipoproteinaemias**

SUCRASE

An intestinal disaccharidase which catalyses the hydrolysis of sucrose into glucose and fructose. A deficiency of the enzyme occurs, along with a deficiency of the other intestinal disaccharidases, in conditions where there is generalised disease of the intestinal wall. A congenital deficiency of sucrase, usually co-existing with isomaltase deficiency, can occur. Acquired sucrase–isomaltase deficiency has also been described but is much rarer.

See also: **disaccharidases and disaccharidase deficiency, sucrose tolerance test**

SUCROSE TOLERANCE TEST

A test which can be used for the diagnosis of intestinal sucrase deficiency. It consists of giving an oral load of sucrose and then measuring the blood level of glucose (resulting from the hydrolysis of sucrose). A failure of the blood glucose levels to rise significantly is suggestive of sucrose malabsorption. However, when this occurs, the test should be followed by a glucose tolerance test to ensure that the failure to absorb sucrose was not due to generalized malabsorption.

See also: **disaccharidases and disaccharidase deficiency**

Further reading: Editorial (1977). Sucrose malabsorption. *Br. Med. J.*, **1**, 1558

SULKOWICH'S TEST

A semi-quantitative test for the estimation of urinary calcium excretion. It consists of adding oxalic acid and ammonium oxalate in acetic acid to the urine and assessing the degree of turbidity of the calcium oxalate precipitate.

SULPHAEMOGLOBIN

Haemoglobin containing an extra sulphur atom. It can be formed by the action of certain drugs, e.g. phenacetin, and can be detected by its characteristic absorption spectrum.

SULPHITE OXIDASE DEFICIENCY

This is a heriditary disorder of sulphur metabolism in which sulphite, thiosulphate and S-sulphocysteine occur in increased amounts in the urine. Affected individuals can have dislocated lenses together with a sudden onset of paresis, and death at an early age results.

SUPPRESSION TESTS

These are a class of *in vivo* tests which are used mainly for diagnosing excessive hormone secretion by a gland. A substance is given which normally suppresses hormone secretion by negative feedback inhibition. Failure to suppress the hormone levels suggests that the secretion is not under feedback control.

The low dose dexamethasone suppression test (qv) is an example of such a test. Dexamethasone is a powerful cortisol analogue which is capable of suppressing ACTH production and therefore cortisol secretion. In Cushing's syndrome, however, the cortisol levels do not fall, e.g. due to pituitary disease (when the feedback mechanism is insensitive) or due to adrenal carcinoma or adenoma when cortisol secretion is autononomous.

See also: **stimulation tests**

SUXAMETHONIUM SENSITIVITY

A condition in which affected individuals have low levels of an abnormal cholinesterase. This results in a prolonged period of apnoea after the administration of suxamethonium, a muscle relaxant normally broken down by cholinesterase.

See also: **cholinesterase**

SWEAT TEST

A test that can be used for the diagnosis of cystic fibrosis (qv). In this condition there is a blockage in pancreatic and bronchial secretions and also disordered sweat gland function. The latter condition results in the production of sweat with high sodium and chloride content.

In order to collect the sweat for sodium or chloride measurement, a drug such as pilocarpine is introduced into the skin by iontophoresis (qv) in order to stimulate sweat production. The sweat is absorbed onto filter paper which has been placed on the skin. It is then eluted off the filter paper by placing the discs in distilled water. This is analysed for sodium or chloride content.

Chloride sensitive, ion-specific electrodes placed on the skin have been used as an alternative technique.

SYMPTOMATIC CUTANEOUS HEPATIC PORPHYRIA

An acquired form of hepatic porphyria which can be caused by severe liver disease or toxins. It is accompanied by an increased urinary excretion of porphyrins, especially uroporphyrins. Hirsutism and cutaneous lesions are among the clinical symptoms.

Further reading: Elder, G.H. (1976). Acquired disorders of haem synthesis. In Marks, V. and Hales, C.N. (eds.) *Essays in Medical Biochemistry*. Vol. 2, p. 75. (London: The Biochemical Society and Association of Clinical Biochemists)

SYNACTHEN STIMULATION TEST

Synacthen (tetracosactrin) is a synthetic peptide having ACTH activity. It is used in the diagnosis of adrenocortical hypofunction. Injection of synacthen should normally result in an increase in plasma cortisol levels after 30 minutes as a result of stimulation of the adrenal gland. An impaired response suggests adrenocortical hypofunction. A further test may then be performed in order to distinguish between primary adrenocortical hypofunction (Addison's disease) and secondary adrenocortical hypofunction. This consists of three injections of synacthen on successive days. An improvement in cortisol production indicates secondary adrenocortical hypofunction. No improvement occurs if there is primary adrenocortical insufficiency.

See also: **Addison's disease**

SYNOVIAL FLUID

The biochemical tests which can be performed on synovial fluid include protein content (which is raised in inflammatory diseases of the joint, e.g. gout or arthritis) and microscopic examination for uric acid cystals (in gout) or calcium pyrophosphate crystals (in pseudogout). Increased levels of certain enzymes (e.g. 5'-nucleotidase) in synovial fluid can be found in rheumatoid arthritis.

Further reading: Scott, J.T. (1975). The analysis of joint fluids. *Br. J. Hosp. Med.*, **14**, 653

SYSTEMIC LUPUS ERYTHEMATOSUS

An autoimmune disease in which antibodies are produced which act against platelets, erythrocytes and other cells. Among these antibodies is an IgG called antinuclear factor which acts against the nuclei of the patients own cells especially leukocytes. Anti-DNA antibodies (qv) can be detected in the serum of many patients with the condition. The disease is confined mainly to young women. Among the clinical findings are arthritis, lung and heart disorders, and hepatomegaly.

Further reading: General list of clinical text books

T

t-TEST

A statistical test for determining if there is a significant difference between the mean values of two sets of data. From the two sets of data, a statistical parameter called *t* can be calculated. From statistical tables, the critical value of *t* for the number of observations can be found. If the calculated value for *t* exceeds the value in the tables, the difference is significant, i.e. unlikely to be due to chance alone.

Further reading: Swinscow, T.D.V. (1978). *Statistics at Square One*. 3rd Edn. (London: British Medical Journal)

T_3, T_4

Abbreviations for tri-iodothyronine (qv) and thyroxine (qv) respectively.

TAKATA-ARA REACTION

A flocculation test for the detection of changes in serum globulin levels. It consists of adding serum to a buffered mercuric chloride solution and assessing the turbidity.

TAMM-HORSFALL PROTEIN

A renally derived mucoprotein which is normally excreted in small amounts (4 mg/day) in the urine.

See also: **proteinuria**

TANGIER DISEASE

A rare genetic disorder in which there is a deficiency of α-lipoproteins in the serum. The accumulation of cholesterol esters in the tissues is a feature of the disease. Hepatomegaly and lymphadenopathy are among the clinical findings.

Further reading: Herbert, P.N., Gotto, A.M. and Fredrickson,

D.S. (1978). Familial lipoprotein deficiency (Abetalipoproteinaemia, hypobetalipoproteinaemia and Tangier disease). In Stanbury, J.B., Wyngaarden, J.B. and Fredrickson, D.S. (eds.) *The Metabolic Basis of Inherited Disease.* 4th Edn., p. 544 (New York: McGraw-Hill)

TANNED RED CELL TECHNIQUES

Many substances will bind to erythrocytes if these cells have been pretreated with tannic acid. The erythrocytes can then be used in haemagglutination techniques (qv).

TARTRATE-LABILE ACID PHOSPHATASE

A term synonymous with prostatic acid phosphatase.

See: **acid phosphatases**

TAY-SACHS DISEASE (GM_2 GANGLIOSIDOSIS)

An inborn error of lipid storage. The most common form of the disease is due to a deficiency of ganglioside GM_2-hexosaminidase, an enzyme which catalyses the breakdown of ganglioside GM_2. As a result, this ganglioside accumulates in nervous tissue. Paralysis, dementia and blindness are features of the condition. Death occurs in early infancy.

Further reading: O'Brien, J.S. (1978). The gangliosidoses. In Stanbury, J.B., Wyngaarden, J.B. and Fredrickson, D.S. (eds.) *The Metabolic Basis of Inherited Disease.* 4th Edn., p. 841. (New York: McGraw-Hill)

TECHNETIUM UPTAKE TEST

Radioactively labelled technetium (^{99m}Tc) can be used instead of radioactive iodine (see **radioactive iodine neck uptake test**) for the assessment of thyroid gland activity. Its half life is six hours and thus the radiation dose to the patient is less than that given by ^{131}I or ^{132}I.

TERATOMA

A tumour that may arise in the ovaries, testes and other sites. One type of tumour, chorion-epithelioma, is associated with grossly elevated secretion of human chorionic gonadotrophin

(HCG). Following removal of the tumour, HCG levels may be measured in order to monitor progress. Serum α-fetoprotein levels may also be increased in teratomas in both sexes.

Further reading: Editorial (1979). Teratomas of the ovary. *Br. Med. J.*, **1**, 1034

TES-TAPE

A reagent strip manufactured by Lilley for the detection of urinary glucose. It is based on the glucose oxidase-peroxidase reaction.

Further reading: Kutter, D. (1977). *Rapid Clinical Diagnostic Tests*. (Munich–Vienna–Baltimore: Urban and Schwarzenberg)

TESTICULAR FEMINIZATION SYNDROME

In this disorder, phenotypic females are discovered to have testicular gonadal tissue due to a male (i.e. XY) chromosomal complement. Testosterone secretion appears to be normal and the condition is thought to be a failure of the peripheral tissues to respond to testosterone. The disorder is thought to have a familial incidence, being transmitted by female carriers.

Further reading: Wilson, J.D. and Macdonald, P.C. (1978). Male pseudohermaphroditism due to androgen resistance: Testicular feminization and related syndromes. In Stanbury, J.B., Wyngaarden, J.B. and Fredrickson, D.S. (eds.) *The Metabolic Basis of Inherited Disease*. 4th Edn., p. 894. (New York: McGraw-Hill)

TESTIS

Both LH and FSH influence the action of the testis. LH stimulates the Leydig (interstitial) cells to produce testosterone, the hormone responsible for the development of the male sexual characteristics. FSH stimulates spermatogenesis by its action on the Sertoli cells.

Biochemical diagnosis of testicular disorders

(1) *Hypogonadism*. This can be a primary defect due to testicular atrophy or secondary to pituitary hypofunction. In both these conditions urinary 17-oxosteroid excretion

and plasma testosterone levels are low. However, in primary but not secondary hypogonadism, elevated FSH and LH values occur. The HCG stimulation test (qv) may also help in distinguishing between primary and secondary hypogonadism.

(2) *Hypergonadism.* This can result from congenital adrenal hyperplasia or an interstitial cell tumour of the testis. Plasma testosterone and urinary 17-oxosteroid excretion are raised in both conditions.

(3) *Carcinoma.* In some cases of testicular carcinoma, high urine levels of human chorionic gonadotrophin can be found. Serum α-fetoprotein levels may also be increased.

(4) *Infertility.* If the male infertility is due to seminiferous tubular failure with a resulting failure of spermatogenesis, hormone analysis is of little use. Testosterone and LH levels can be normal, as is the response to HCG stimulation. FSH levels however may be high. Analysis of the semen may be of more use. (see also **seminal fluid fructose)**

Further reading: Anderson, D.C. (1978). Endocrine function of the testis. In O'Riordan, J.L.H. (ed.) *Recent Advances in Endocrinology and Metabolism.* No. 1, p. 111. (Edinburgh, London and New York: Churchill Livingstone)

Hall, R., Anderson, J., Smart, G.A. and Besser, M. (1974). *Fundamentals of Clinical Endocrinology.* 2nd Edn. (London: Pitman Medical Publishing Co.)

Editorial (1979). Patterns of male infertility. *Br. Med. J.*, **2**, 1169

TESTOSTERONE

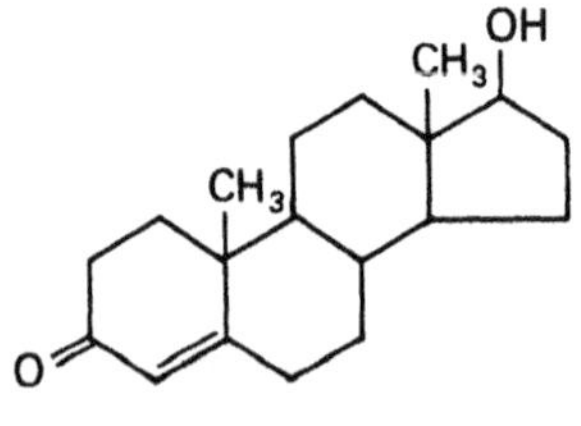

An androgenic steroid hormone, secreted by the Leydig (interstitial) cells of the testis in response to LH. A small amount of testosterone is also formed from dehydroepiandrosterone, synthesized by the adrenals. In women, some results from the peripheral conversion of the ovarian androgen, androstenedione, to testosterone. It is carried in the blood bound to sex hormone binding globulin. Testosterone is reduced by a number of tissues to dihydrotestosterone which is an even more potent androgen than testosterone itself. Testosterone is responsible for the development of the male sexual characteristics.

Decreases in plasma testosterone levels are found in both primary and secondary hypogonadism. In women, high plasma levels can be found in virilizing tumours of the ovaries or adrenals.

The hormone is usually measured by radioimmunoassay.

TETRACOSACTRIN STIMULATION TEST

See: **synacthen stimulation test**

TETRAHYDROCORTISOL

One of the major urinary excretion products of cortisol. It is one of the compounds of the 17-oxogenic steroid and 17-hydroxycorticosteroid group.

TETRAHYDROCORTISONE

One of the major urinary excretion products of cortisol. It is one of the compounds of the 17-oxogenic steroid and 17-hydroxycorticosteroid group.

THALASSAEMIAS

A group of hereditary disorders, found mainly in Mediterranean peoples, in which a haemoglobin molecule containing abnormal proportions of subunits is produced. The normal haemoglobin molecule consists of two α and two β chains. However, in α-thalassaemias, there is a deficient production of α chains and haemoglobin molecules are produced which are deficient in α chains. Likewise in β-thalassaemia, there is a deficient production of β chains, and haemoglobin molecules are produced which are deficient in β chains. The abnormal haemoglobins can be identified by electrophoresis.

Thalassaemia major presents in early childhood with hepatosplenomegaly, anaemia and jaundice. Thalassaemia minor usually presents with a milder form of anaemia. Thalassaemia major has been referred to as the homozygous state and thalassaemia minor as the heterozygous (carrier) state but this is probably not true in many cases of the disease.

Further reading: Weatherall, D.J. (1978). The thalassaemias. In Stanbury, J.B. Wyngaarden, J.B. and Fredrickson, D.S. (eds.) *The Metabolic Basis of Inherited Disease*. 4th Edn., p. 1508. (New York: McGraw-Hill)

Huisman, T.H.J. (1972). Normal and abnormal human haemoglobins. In Bodansky, O. and Latner, A.L. (eds.) *Advances in Clinical Chemistry*. Vol. 15, p. 150. (New York: Academic Press)

THALLIUM

Thallium salts are sometimes used as rodenticides. Their ingestion by humans can result in toxic symptoms which include loss of hair and nails. It can be detected in urine by a screening test which involves its oxidation and complexation with methyl-violet to form a blue compound which can be extracted into benzene.

THIAMINE

A water-soluble vitamin of the B complex, obtained in the diet from plant sources. It is a component of thiamine pyrophosphate, a cofactor required for the enzyme, transketolase (qv), and also for the decarboxylation of α-oxoacids (e.g. the conversion of pyruvate to acetyl CoA).

A deficiency of thiamine causes beri-beri, features of which include emaciation, cardiomyopathy and neurological disturbances.

Biochemical diagnosis of thiamine deficiency

(1) It can be estimated fluorimetrically in urine after its oxidation to thiochrome by alkaline ferricyanide.

(2) Blood pyruvate levels may be increased.

(3) The erythrocyte transketolase level may be low. This is increased by the addition of thiamine pyrophosphate.

Further reading: General list of analytical and clinical textbooks

THIN-LAYER CHROMATOGRAPHY

A form of chromatography in which the chromatographic material (silica, cellulose, Sephadex) is thinly coated onto a plate of backing material, e.g. glass, plastic or aluminium foil. Passage of the solvent up (or, in the case of thin layer Sephadex chromatography, down) the plate results in the separation of a mixture of substances and these can be located by specific sprays. Thus depending on the chromatographic material and the solvents, thin-layer chromatography can be used for adsorption, partition or gel filtration chromatography.

Thin-layer chromatography can be used for the separation of drugs, amino acids, sugars and catecholamine metabolites, and for the determination of lecithin–sphingomyelin ratios.

Further reading: General list of analytical textbooks

THORMÄHLENS TEST

A test for the detection of melanogen in urine, based on its reaction with sodium nitroprusside to give a greenish-blue colour in acid solution.

THORPE AND STONE CLASSIFICATION

A means of classifying the hyperlipoproteinaemias, based on the size of the lipoprotein molecules as assessed by membrane filtration and nephelometry. The hyperlipoproteinaemias are expressed in terms of an SML profile where S, M and L stand for small, medium and large particles respectively.

The S particles are equivalent to β (low density) lipoproteins.
The M particles are equivalent to pre-β (very low density) lipoproteins.
The L particles are equivalent to chylomicrons.

THROMBIN

A protein which converts fibrinogen to fibrin, the fibrous material of the blood clot. Thrombin itself is formed from prothrombin by the action of thromboplastin and calcium.

THROMBOPLASTIN

A blood clotting factor which, along with calcium, converts prothrombin to thrombin.

THYMOL TURBIDITY

A flocculation test which consists of adding serum to a buffered solution of thymol. The degree of turbidity is proportional to the increase in the γ-globulin level. The reaction is enhanced by increases in the α- and β-globulin levels and by a low serum albumin concentration.

THYROCALCITONIN

See: **calcitonin**

THYROGLOBULIN

A large protein molecule of the thyroid gland into which are incorporated the thyroid hormones and their precursors. The hormones are released into the circulation by breakdown of thyroglobulin by proteolytic enzymes.

THYROID AUTOANTIBODIES

A number of circulating thyroid autoantibodies have been described. They include;

(1) An antibody acting against the thyroid microsomal fraction. It can be detected by complement fixation tests or immunofluorescence.

(2) An antibody acting against thyroglobulin (the protein which acts as a store of thyroid hormones in the thyroid colloid). These antibodies can be detected by double diffusion tests, immunofluorescence or agglutination tests.

(3) Another antibody acting against thyroid colloid has also been described. This can also be detected by immunofluorescence.

These three antibodies acting against different thyroid constituents are associated with destructive inflammatory lesions of the thyroid. This results in disturbed thyroid function which may present as either hypo or hyperthyroidism.

In addition to these three autoantibodies, there is a protein known as long-acting thyroid stimulator (LATS). This can also be considered as a thyroid autoantibody since it is an immunoglobulin which causes overstimulation of the thyroid gland, possibly acting at the site of pituitary TSH activity. LATS is found in the serum of many patients with hyperthyroidism. It can be measured by bioassay.

Further reading: General list of clinical textbooks

THYROID FUNCTION TESTS

The biochemical tests that can assist in the diagnosis of thyroid disease are:

(1) Serum total thyroxine. High levels suggest hyperthyroidism; low levels suggest hypothyroidism.

(2) Measurement of the free (physiologically active) thyroxine in serum. This may be determined directly or can be derived (the free thyroxine index) if a T_3 resin uptake or a thyroxine-binding globulin level is measured.

(3) Serum tri-iodothyronine. This is only useful for the diagnosis of hyperthyroidism.

(4) Serum TSH. This is low in hyperthyroidism and secondary hypothroidism and high in primary hypothyroidism.

(5) TRH stimulation test. This can be used to find the cause of hypothyroidism.

(6) Thyroid autoantibodies can be found in both hypo- and hyperthyroidism.

(7) Radioactive iodine neck uptake test. High uptakes suggest hyperthyroidism; low uptakes suggest hypothyroidism. This test may be followed by a TSH stimulation test or a T_3 suppression test.

(8) There are also a number of incidental tests which can be performed, e.g. cholesterol, which can be low in hypothyroidism, or serum calcium, which can be raised in hyperthyroidism.

See separate entries for all of these tests.

Further reading: Wellby, M.L. (1976). The laboratory diagnosis of thyroid disorders. In Bodansky, O. and Latner, A.L. (eds.) *Advances in Clinical Chemistry*. Vol. 18, p. 104. (New York: Academic Press)

THYROID-STIMULATING HORMONE (TSH, THYROTROPHIN)

A glycoprotein hormone secreted by the pituitary in response to a hypothalamic releasing factor (TRF). High levels of thyroxine inhibit TSH release and this is thought to be due to thyroxine blocking the response of the pituitary to TRF. TSH acts by stimulating the production of thyroid hormones from the thyroid gland. High serum levels are found in primary hypothyroidism while low levels occur in hyperthyroidism and secondary hypothyroidism. TSH is usually measured by radioimmunoassay.

THYROTOXICOSIS

See: **hyperthyroidism**

THYROTROPHIN

See: **thyroid-stimulating hormone (TSH)**

THYROTROPHIN-RELEASING FACTOR (TRF, THYROTROPHIN-RELEASING HORMONE, TRH)

A tripeptide, released from the hypothalamus, which is capable of stimulating TSH release from the pituitary. The factors which control TRF secretion are not well understood. In infants, but not adults, TRF is secreted on exposure to cold. The negative feedback inhibition by thyroxine of its own synthesis is believed to occur at the pituitary level where it may block the response of this gland to TRF, but it is also possible that thyroxine may inhibit TRF release.

Further reading: Hall, R. and Gomez-Pan, A. (1976). The hypothalamic regulatory hormones and their clinical applications. In Bodansky, O. and Latner, A.L. (eds.) *Advances in Clinical Chemistry*. Vol. 18, p. 174. (New York: Academic Press)

THYROXINE (T_4)

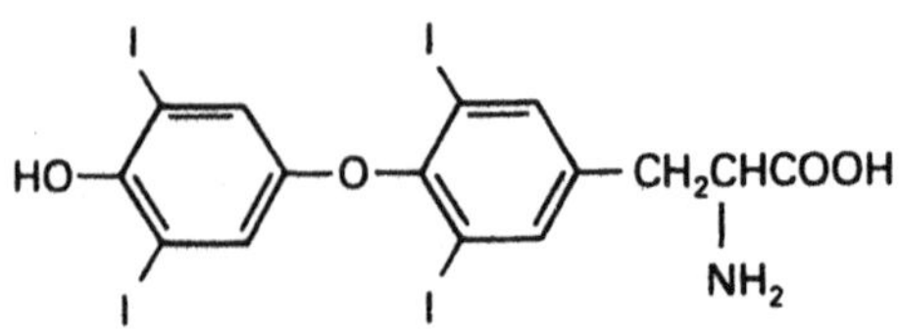

The major hormone secreted by the thyroid gland. It influences the rate of many of the metabolic reactions of the body and is required for normal mental development and growth.

Biosynthesis

(1) Iodide is taken up by the thyroid gland.

(2) The trapped iodide is converted to iodine.

(3) The iodine is attached to tyrosine to form mono-iodotyrosine (MIT) and di-iodotyrosine (DIT).

(4) MIT and DIT are coupled to form tri-iodothyronine (one molecule of MIT and one molecule of DIT combined together) and thyroxine (two molecules of DIT combined together).

Thyroxine (and tri-iodothyronine) are synthesized in the thyroid protein thyroglobulin. They are released into the circulation by the action of proteolytic enzymes.

Control of thyroxine secretion

The synthesis and secretion of thyroxine is promoted by the pituitary hormone, thyroid-stimulating hormone (TSH). TSH secretion, in turn, is dependent on the secretion of thyrotrophin-releasing factor (TRF) from the hypothalamus. High levels of thyroxine inhibit TSH secretion by negative feedback to the pituitary.

Serum thyroxine

The majority of the thyroxine in the blood is carried on plasma proteins, the main ones being thyroxine-binding globulin and thyroxine-binding pre-albumin. It is, however, the small fraction of the free hormone which is physiologically active.

Thyroxine catabolism

(1) Some thyroxine is conjugated by the liver and excreted in the bile.

(2) Deiodination of thyroxine to tri-iodothyronine by the tissues occurs in man. The deiodination is random and results in the formation of 3,3′,5-tri-iodothyronine (T_3) and 3,3′,5′-tri-iodothyronine (reverse T_3)

(3) A further means of thyroxine removal is by deamination and decarboxylation, resulting in the formation of the breakdown products tetra-iodothyroacetic acid (TETRAC) and tri-iodothyroacetic acid (TRIAC)

Disorders

Increased circulating thyroxine causes hyperthyroidism while decreased secretion results in hypothyroidism. See separate entries for these two conditions.

Measurement of serum thyroxine

(1) Thyroid hormone activity can be determined indirectly as the basal metabolic rate (q.v.).

(2) Thyroxine can be measured as protein-bound iodine (PBI). This measures both thyroxine and the smaller proportions of T_3. The method involves four basic steps;

(a) Precipitation of the serum proteins. More than 90% of the serum thyroxine and T_3 is protein bound and is precipitated in this step.

(b) Washing of the precipitate to remove the trapped inorganic iodide.

(c) A digestion step in which the thyroid hormones in the precipitate are oxidized to liberate free iodine.

(d) The iodine is measured by its ability to catalyse the reduction by arsenite of the yellow coloured ceric ion to the colourless cerous ion.

The entire procedure can be automated on continuous flow instruments, with the use of a digestor. The main disadvantage of protein-bound iodine determinations is interference from iodine-containing compounds such as drugs and X-ray contrast media.

(3) Butanol-extractable iodine (BEI) methods. These were regarded as an improvement on the PBI, although they are now little used because they are difficult to perform and imprecise. They consist of extracting the serum with n-butanol. This extracts thyroxine, T_3, mono-iodotyrosine, di-iodotyrosine and inorganic iodide. The butanol extract is washed with alkali which removes mono and di-iodotyrosines and inorganic iodide. The butanol extract, which now contains mostly thyroxine and T_3, can be

analysed for iodine content as in the last step of PBI determination.

(4) Column methods. Thyroxine can be isolated by ion-exchange column chromatography and its iodine content determined as described for PBI.

(5) Competitive protein binding techniques (q.v.). In many methods this consists of the following steps:

  (a) Extraction of the serum thyroxine with a suitable solvent, e.g. ethanol.

  (b) An aliquot of the extract is added to a mixture of thyroid-binding globulin (TBG) and radioactively labelled T_4. This mixture is equilibrated, during which time the unlabelled (i.e. patient's) thyroxine displaces some of the labelled thyroxine from the TBG.

  (c) The unbound thyroxine is removed by a resin or other means and the radioactivity either in this or in the supernatant is counted. From this, the level of the patient's serum thyroxine can be calculated.

(6) Radioimmunoassay methods. Specific thyroxine antisera have been raised and can be used to measure serum thyroxine by radioimmunoassay. Most laboratories now probably measure thyroxine by this means. Radioimmunoassays have the following advantages over the competitive protein binding techniques:

  (a) They can be performed directly on sera without a prior extraction.

  (b) They have fewer stages.

  (c) They are less susceptible to interference by drugs.

  Many commercial kits are available for measurement of thyroxine by RIA. These employ a variety of different principles to separate the bound and free antigen. For instance, the antibody may be incorporated into a plastic tube, or attached to cellulose or glass granules, in which case the bound and free antigen can be separated by centrifugation. In another technique, the antibody is attached to magnetic particles, enabling separation by magnetic means.

(7) Several enzyme-immunoassay kits are also commercially available.

Free thyroxine

As previously mentioned, this is the portion of the total serum thyroxine which is physiologically active. An indication of the free thyroxine level is therefore desirable when investigating thyroid function. There are several approaches to its determination:

(1) Free thyroxine index. This can be calculated from the total serum thyroxine and either the T_3 uptake value (q.v.) or the thyroxine-binding globulin level.

(2) Dialysis techniques. These are not suitable for routine use however.

(3) Commercial kits (e.g. the Corning Immophase) are available which measure the free thyroxine directly. The basis of these methods is measurement of the rate at which thyroxine binds onto immobilized antibody.

Further reading: General list of analytical and clinical textbooks

THYROXINE BINDING GLOBULIN (TBG)

The plasma protein which binds the majority of the circulating thyroxine. It also binds tri-iodothyronine, although less avidly than thyroxine. In normal subjects, it is about 25% saturated. Decreased TBG levels are found in conditions where there is a generalized hypoproteinaemia, such as nephrotic syndrome. Cases of congenital TBG deficiency have also been described. Increased serum TBG levels are found in pregnancy and in patients on oral contraceptives.

Measurement

(1) TBG can be determined by immunochemical methods, e.g. by the Laurell 'rocket' technique. If the total serum thyroxine in the sample is known, the amount of the free hormone can be calculated.

(2) The T_3 uptake test (q.v.) gives an indication of the degree of saturation of the TBG.

THYROXINE-BINDING PRE-ALBUMIN (TBPA)

A plasma protein which binds about 30% of the circulating plasma thyroxine. Unlike thyroxine-binding globulin, however, it does not bind tri-iodothyronine.

TITRATABLE ACIDITY

This is a measure of the renal excretion of hydrogen ions. It can be determined by measuring the amount of alkali required to titrate a fixed volume of the urine to pH 7.4. Titratable acidity is increased when acid-forming foods have been taken and in some acidotic conditions, such as diabetic ketoacidosis when keto acids are excreted in the urine. Titratable acidity can also be measured in the ammonium chloride loading test (q.v.), a procedure designed to test the ability of the kidneys to excrete an acid load.

TOCOPHEROLS

See: **vitamin E**

TOLBUTAMIDE TEST

A test that can be used for the diagnosis of insulinoma. Tolbutamide is a drug which stimulates insulin release from the pancreas. It is injected intravenously and blood samples are collected for up to three hours. In normal subjects, the blood glucose level falls and then returns to normal. However, in patients with insulinoma, a greater fall can occur and the hypoglycaemia persists for much longer.

See also: **insulinoma**

TOLLEN'S TEST

A test for glucuronides in urine, based on their reaction with naphthoresorcinol to give a red colour.

TONOMETRY

The Astrup technique (q.v.) is an example of a tonometric method. In this procedure, a blood sample is brought into equilibrium with a gas phase.

TOTAL CARBON DIOXIDE

See: **bicarbonate**

TOTAL IRON BINDING CAPACITY

See: **iron binding capacity**

TOTAL PROTEIN

See: **proteins**

TRACE ELEMENT ANALYSIS

In clinical chemistry, the techniques that can be used for trace element analysis include:

(1) Chemical methods, i.e. the reaction of the element with a chemical to produce a coloured compound which can be estimated colorimetrically.

(2) Atomic absorption spectrophotometry, especially the flameless variety.

(3) Anodic stripping voltametry.

(4) Ion-selective electrodes.

See separate entries for these techniques.

Further reading: Special issue on trace elements in clinical chemistry. *Clin. Chem.*, (1975). **21**, No. 4

Delves, H.T. (1976). The clinical value of trace-metal measurements. In Marks, V. and Hales, C.N. (eds.) *Essays in Medical Biochemistry*. Vol. 2, p. 37. (London: The Biochemical Society and Association of Clinical Biochemists)

TRANSCOBALAMINS

Vitamin B_{12} is carried in the plasma, bound to two globulins, transcobalamins I and II.

TRANSCORTIN (CORTICOSTEROID-BINDING GLOBULIN)

An α-globulin which binds cortisol in the plasma. Increased plasma transcortin levels are found in pregnancy and in women on oral contraceptives. The active portion of the plasma cortisol is that which is unbound; cf. other protein bound substances such as calcium or thyroxine.

Transcortin can be used in competitive protein binding assays for the estimation of cortisol.

TRANSFERRIN

The protein on which the bulk of the plasma iron is transported. Transferrin (formerly called siderophilin) has a molecular weight of 80000 and migrates electrophoretically as a β_1-globulin. Each molecule of transferrin can combine with two ferric ions but in a normal person the serum transferrin saturation is only about 25–30%. Increased serum transferrin levels are found in iron deficiency anaemia and pregnancy. Low levels are found in conditions in which a generalized hypoproteinaemia is found, e.g. malnutrition, nephrotic syndrome and cirrhosis.

Measurement

Transferrin may be determined directly using immunochemical methods such as radial immunodiffusion or crossed electrophoresis. It can be measured indirectly by determining the amount of iron it can bind (see **iron binding capacity**).

Further reading: Dixon K. (1973). Technical Bulletin No. 28. Routine clinical measurements of transferrin in human serum. *Ann. Clin. Biochem.*, **10**, 127

TRANSKETOLASE

An enzyme found in many tissues (including red cells) which catalyses the following two reactions in the pentose phosphate pathway:

$$\text{xylulose-5-phosphate} + \text{ribose-5-phosphate} \rightleftharpoons \text{sedheptulose-7-phosphate} + \text{glyceraldehyde-3-phosphate}$$

$$\text{xylulose-5-phosphate} + \text{erythrose-4-phosphate} \rightleftharpoons \text{fructose-6-phosphate} + \text{glyceraldehyde-3-phosphate}$$

Thiamine pyrophosphate is required as a cofactor for transketolase activity. Decreased erythrocyte transketolase activity is therefore found in thiamine deficiency and activity is increased by the addition of thiamine pyrophosphate to the assay system.

Measurement

Transketolase activity can be measured by:

(1) Measuring the rate of disappearance of ribose-5-phosphate in the first reaction with orcinol reagent (i.e. Bial's test, q.v.).

(2) Measuring the fructose formed in the second reaction by anthrone reagent.

(3) A coupled enzyme system can also be used. The glyceraldehyde-3-phosphate formed is converted to dihydroxyacetone phosphate by triosephosphate isomerase and this is followed by reduction by glycerol phosphate dehydrogenase. The oxidation of NADH in this last stage is measured spectrophotometrically.

Further reading: Wilkinson, J.H. (1976). *The Principles and Practice of Diagnostic Enzymology*. (London: Edward Arnold)

TRANSMISSION *(T)*

The proportion of light passing through an absorbing material. It is expressed as the ratio of the emergent light to the incident light (for example as the percentage transmission). Transmission varies inversely with absorption (A) as follows:

$$A = -\log_{10} T$$

TRF, TRH

See: **thyrotrophin releasing factor**

TRH STIMULATION TEST

A test that can be used for the differential diagnosis of hypothyroidism. It consists of intravenously injecting TRH and then measuring the subsequent response of TSH levels. A normal response indicates hypothalamic disease (assuming the patient is hypothyroid). An absent response indicates pituitary disease while an exaggerated response occurs in primary hypothyroidism.

Further reading: General list of clinical textbooks

TRICYCLIC ANTIDEPRESSANTS

A group of drugs used in the treatment of depression. They include desipramine, imipramine, amitriptyline and nortriptyline. Overdoses are frequently encountered and, in such cases, the drug can be detected in urine or gastric contents by a number of screening tests.

TRIGLYCERIDES

These are lipids formed by the condensation of glycerol with fatty acids. They can be synthesized in a number of different ways:

(1) They can be synthesized by the gut, following the absorption of hydrolyzed dietary lipid. They are combined with small amounts of protein, cholesterol and phospholipid to form chylomicrons. These are removed from the blood, mainly by adipose tissue. An enzyme, lipoprotein lipase is responsible for this hydrolysis of the chylomicron triglycerides. The free fatty acids liberated by these processes are resynthesized into triglycerides by the adipose tissue and stored.

(2) Endogenous liver triglyceride synthesis can occur in conditions when there is an excess of free fatty acids reaching the liver, e.g. diabetes, or when there is excessive hepatic *de novo* free fatty acid synthesis. Triglycerides synthesized in this way are incorporated into pre-β-lipoproteins.

Increased serum triglycerides

This can occur in diabetes mellitus, nephrotic syndrome and in types I, IIb, III, IV and V hyperlipoproteinaemias.

Measurement

(1) *General kinase methods.* The triglycerides are hydrolysed by either alkali or lipase. The glycerol formed as a result can be estimated using the following coupled reaction system:

$$\text{glycerol} + \text{ATP} \xrightarrow{\text{glycerol kinase}} \text{glycerol-3-phosphate} + \text{ADP} \quad (1)$$

$$\text{ADP} + \text{phosphoenolpyruvate} \xrightarrow{\text{pyruvate kinase}} \text{ATP} + \text{pyruvate} \quad (2)$$

$$\text{pyruvate} + \text{NADH} \xrightarrow{\text{lactate dehydrogenase}} \text{lactate} + \text{NAD} \quad (3)$$

The reaction is therefore followed by measuring the decrease in absorption of the NADH used in the last reaction.

Another approach is to convert the glycerol-3-phosphate, formed in reaction (1), to dihydroxyacetone phosphate using the enzyme, glycerol phosphate dehydrogenase.

glycerol-3-phosphate + NAD

⇅ glycerol phosphate dehydrogenase

dihydroxyacetone phosphate + NADH

This reaction is followed by measuring the increase in absorption of the NADH formed.

In these and other methods a correction may be applied for the small amount of free glycerol present in the sample. This is achieved by running a blank.

(2) *Colorimetric methods.* The triglycerides are initially separated from phospholipids and other interfering substances by either column chromatography (e.g. Zeolite) or solvent extraction. The isolated triglycerides are hydrolysed by alkali and the liberated glycerol is oxidized to formaldehyde by periodate. The formaldehyde can be estimated in one of two ways.
   (a) By its reaction with chromotropic acid in sulphuric acid to give a pink coloured compound.
   (b) By its combination with ammonium ion and acetylacetone to give a yellow coloured lutidine derivative (the Hantzsch reaction). This compound can be measured either colorimetrically or fluorimetrically. Both the colorimetric and fluorimetric methods can be used on continuous flow instruments.

Further reading: General list of analytical and clinical textbooks

TRI-IODOTHYRONINE (T_3)

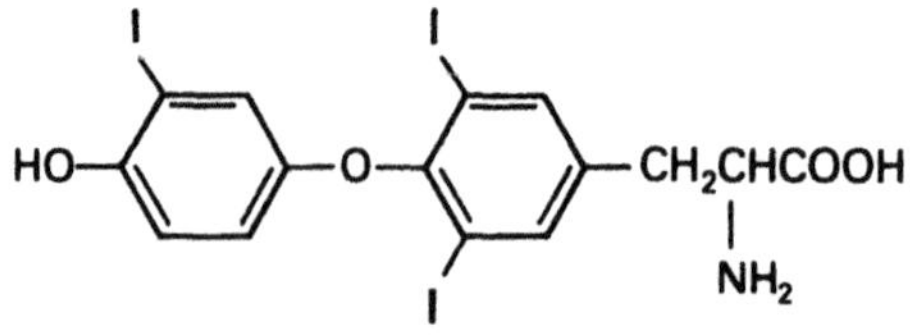

T_3 is found in the serum as a result of secretion by the thyroid gland (where it is synthesized) and as a result of the degradation of circulating thyroxine.

It is metabolically more active than thyroxine. It represents 1–2% of the total thyroid hormones in the serum and is carried by thyroxine-binding globulin, although it has a lower affinity for this protein than thyroxine.

It is usually measured by radioimmunoassay. Low serum levels are found in hypothyroidism while high levels occur in hyperthyroidism. Occasionally a raised serum T_3 level may be found in the presence of a normal level of serum thyroxine ('T_3 toxicosis'). This can occur when there is dietary iodine deficiency. It may also represent an early stage in the development of full thyrotoxicosis.

see also: **reverse T_3**

TRIOLEIN/OLEIC ACID ABSORPTION TEST

A test used for determining the cause of malabsorption. It consists of giving an oral dose of ^{131}I-labelled triolein (a triglyceride consisting of glycerol and oleic acid) and ^{125}I-labelled oleic acid and measuring the radioactivity of each isotope in the blood serum. A decreased absorption of triolein with normal oleic acid absorption suggests impaired pancreatic enzyme secretion. Failure to absorb both triolein and oleic acid suggests the malabsorption is not due to pancreatic dysfunction.

TRYPSIN

A digestive enzyme, secreted by the pancreas, which hydrolyses peptide bonds formed from the carboxyl group of lysine or arginine. It is secreted into the duodenum as its inactive precursor, trypsinogen, and this is converted to trypsin by the action of the intestinal enzyme, enterokinase. The newly formed trypsin can also activate trypsinogen.

Clinical significance

Trypsin can be measured in the duodenal juice or faeces of children suspected of having cystic fibrosis. In this condition there is deficient secretion of trypsin by the pancreas.

Trypsin can also be measured in duodenal juice after pancreatic stimulation by secretin–pancreozymin or by a test meal, e.g. as in the Lundh test (q.v.). A deficient output of trypsin is suggestive of pancreatic disease.

Measurement

(1) Trypsin can be measured semiquantitatively by its ability to hydrolyse the gelatin on the surface of X-ray film. A number of serial dilutions of the specimen are made and these are spotted onto unexposed X-ray film. Gelatin hydrolysis results in the formation of clear areas on the film, the greatest dilution of the film giving a clear area being an indication of the amount of trypsin present.

(2) Trypsin can also be measured by its hydrolysis of synthetic peptide substrates, e.g. benzoyl-arginine ethyl ester. The reaction can be followed by either measuring the increase in absorbance of the reaction product or the change in pH resulting from the liberation of hydrogen ions.

Further reading: Wilkinson, J.H. (1976). *The Principles and Practice of Diagnostic Enzymology.* (London: Edward Arnold)

TRYPTOPHAN

An essential amino acid. In addition to its role as a constituent of proteins, it is involved in the following:

(1) It can be converted by intestinal flora and probably by the tissues to the vitamin, nicotinamide. A dietary deficiency of tryptophan can therefore result in nicotinamide deficiency if dietary intake of nicotinamide is also low. Symptoms resembling those of nicotinamide deficiency can also occur in the inborn error, Hartnup disease, where body tryptophan levels are reduced due to diminished reabsorption of the amino acid from the renal tubules.

(2) Tryptophan is also required for the synthesis of the hormones serotonin and melatonin.

(3) Indican (q.v.) is produced from tryptophan by the action of gut bacteria.

Further reading: Editorial (1974). Studies on tryptophan metabolism in man. *Scand. J. Clin. Lab. Invest.*, **33,** Supplement 136

TRYPTOPHAN LOADING TEST

A test that can be used for the diagnosis of pyridoxine deficiency. It consists of giving an oral load of tryptophan and measuring the excretion of xanthurenic acid. This is the excretion product of 3-hydroxykynurenic acid, an intermediate in the conversion of tryptophan to nicotinic acid. Pyridoxal phosphate is required as a cofactor for the enzyme kynureninase which catalyses the conversion of 3-hydroxykynurenic acid to 3-hydroxyanthrinilic acid. In patients with pyridoxine deficiency, 3-hydroxykynurenic acid accumulates, resulting in the excessive excretion of xanthurenic acid.

TSH

See: **thyroid-stimulating hormone**

TSH STIMULATION TEST

A test used in the differential diagnosis of hypothyroidism when a low radioactive iodine neck uptake has been found. TSH is given to the patient daily and the neck uptake test is repeated. This should result in an increased uptake in normal individuals. No response occurs in primary hypothyroidism. If the hypothyroidism is secondary to pituitary dysfunction, the increase in uptake to normal values may take several days.

See also: **radioactive iodine neck uptake test**

T_3 SUPPRESSION TEST

A high radioactive iodine neck uptake test (q.v.) is usually indicative of hyperthyroidism, although a high uptake can occur if the patient is iodine deficient or is taking anti-thyroid drugs. The neck uptake test can therefore be repeated after tri-iodothyronine (T_3) administration. This normally suppresses TSH secretion, resulting in a low uptake. However, in hyperthyroidism but not in the other conditions listed above, no such suppression occurs.

T_3 THYROTOXICOSIS

A condition in which a patient has a normal serum thyroxine level but an increased level of serum tri-iodothyronine. This can occur when there is iodine deficiency. It may also represent a stage in the development of full thyrotoxicosis.

Further reading: Hollander, C.S. and Shenkman, L. (1972). T_3 toxicosis. *Br. J. Hosp. Med.,* **8,** 393

TUBELESS GASTRIC ANALYSIS

A means of determining whether or not the stomach is secreting adequate amounts of gastric acid, but which does not involve having to pass a tube down into the stomach in order to collect gastric secretions. The Diagnex Blue test (q.v.) is an example of such a test.

TUBULAR SECRETORY CAPACITY

Certain substances are eliminated from the body by secretion from the renal tubules. The rate at which these substances are eliminated is a measure of the tubular secretory capacity. Phenolsulphonephthalein (PSP) or *p*-aminohippuric acid at high levels can be used for this purpose.

T_3 UPTAKE TEST

An *in vitro* test for determining the degree of unsaturation of the serum thyroxine-binding globulin (TBG). Several commercial kits are available for its determination. An excess of radioactive tri-iodothyronine (T_3) is mixed with the patient's serum and this binds to the unoccupied binding sites on the TBG. (The affinity of T_3 for the binding sites is less than that of thyroxine and therefore it does not significantly displace the thyroxine already bound to the protein.) The excess (unbound) radioactive T_3 can be removed by charcoal, red cells, an ion-exchange resin or Sephadex. Radioactive measurement is then made in one of two ways:

(1) The radioactivity removed by the binder (charcoal etc.) may be counted, or

(2) The portion of the radioactivity in solution, i.e. bound to TBG, is measured.

Confusion may arise because in different kits, different forms of counting may be employed. For instance, in those kits which employ counting method (1), a low count indicates a high T_3 uptake by TBG, i.e. an increased number of free binding sites (e.g. hypothyroidism). A high count in the binding fraction indicates a low T_3 uptake by TBG, i.e. a decreased number of free binding sites. This is found in hyperthyroidism.

Alternatively in those kits which employ counting method (2) (e.g. the Thyopac-3 test, Radiochemical Centre, Amersham), a low count indicates a low uptake by the TBG (as in hyperthyroidism). A high count indicates a high uptake by the TBG (as in hypothyroidism).

If the total thyroxine in the patients serum has been determined at the same time, the free thyroxine index (FTI) can be determined. This gives an indication of the free (metabolically active) levels of the hormone in the serum. The FTI is given by the formulae:

$$\text{FTI} = \text{total thyroxine} \times \text{T}_3 \text{ uptake by the binder}$$

or alternatively:

$$\text{FTI} = \frac{\text{total thyroxine}}{\text{T}_3 \text{ uptake by TBG}} \quad \text{(e.g. Thyopac-3)}$$

Uptake measurements are useful when abnormal TBG levels may be suspected in a serum sample, e.g. in pregnant patients, patients on the contraceptive pill, or in hypoproteinaemic states.

Further reading: General list of analytical textbooks

TURBIDIMETRY

A technique for the measurement of the turbidity of solutions. When incident light falls upon a turbid solution, the particles scatter some of the light, and the resultant decreased transmission can be measured by a conventional spectrophotometer. The technique is therefore similar in practice to absorptiometry, although, in this latter technique, the decreased transmission of the light is due to absorption by the molecules in solution, rather than light scattering.

Turbidimetry can be used for the measurement of plasma proteins by their reaction with antisera to form light scattering immune precipitates. Many qualitative and semi quantitative tests are also used in clinical chemistry, e.g. the flocculation tests.

TWO-DIMENSIONAL IMMUNOELECTROPHORESIS

See: **crossed electrophoresis**

TYROSINAEMIA

A rare inborn error of metabolism in which there is abnormal tyrosine and methionine metabolism. The exact enzyme defect is not known and in fact it may not even be an enzyme of tyrosine and methionine metabolism.

Further reading: LaDu, B.N. and Gjessing, L.R. (1978). Tyrosinosis and tyrosinaemia. In Stanbury, J.B., Wyngaarden, J.B. and Fredrickson, D.S. (eds.) *The Metabolic Basis of Inherited Disease.* 4th Edn., p. 256. (New York: McGraw-Hill)

TYROSINASE

An enzyme involved in the conversion of tyrosine to melanin. One form of albinism is due to a deficiency of tyrosinase in melanocytes.

Another tyrosinase is involved in the biosynthesis of adrenaline but this is a different enzyme from that involved in melanin formation.

See also: **albinism**

TYROSINE

An aromatic amino acid which is a precursor of melanin, catecholamines and thyroxine. It is derived both from the diet and the hydroxylation of phenylalanine. Excess tyrosine excretion occurs in the two inborn errors of metabolism, tyrosinosis and tyrosinaemia.

Excess tyrosine in the serum and urine can be detected by chromatography. It can be measured fluorimetrically in serum by its reaction with α-nitroso-β-naphthol and nitrite in the presence of nitric acid. Tyrosine can be detected in urine by the Millon reaction when it gives a red colour with mercuric nitrate in nitric acid containing a trace of nitrous acid.

See also: **tyrosinaemia, tyrosinosis**

TYROSINOSIS

A rare inborn error of metabolism in which there is excessive urinary excretion of tyrosine and *p*-hydroxyphenylpyruvic acid, an intermediate in the catabolism of tyrosine. The enzyme thought to be deficient is *p*-hydroxyphenylpyruvic acid oxidase, the enzyme which converts *p*-hydroxyphenylpyruvic acid to homogentisic acid. Severe liver disease is one of the features of the condition.

Further reading: LaDu, B.N. and Gjessing, L.R. (1978). Tyrosinosis and tyrosinaemia. In Stanbury, J.B., Wyngaarden, J.B. and Fredrickson, D.S. (eds.) *The Metabolic Basis of Inherited Disease*. 4th Edn., p. 256. (New York: McGraw-Hill)

U

UNSATURATED IRON BINDING CAPACITY

See: **iron binding capacity**

URASTRAT

A strip test manufactured by Warner for the estimation of blood urea. It is based on the action of urease on urea. The ammonia formed as a result changes the colour of an indicator dye.

Further reading: Kutter, D. (1977). *Rapid Clinical Diagnostic Tests.* (Munich–Vienna–Baltimore: Urban and Schwarzenberg)

UREA

Urea is the end product of protein and amino acid catabolism. It is filtered through the glomerulus and excreted in the urine, although a portion is reabsorbed by passive diffusion in the renal tubules. Thus measurement of its blood level is an indicator of renal function.

Elevated blood urea levels

(1) Due to faulty excretion, as occurs in renal failure.

(2) Due to increased production. This occurs in increased body protein breakdown (e.g. fever) or as a result of high protein diets.

Decreased blood urea levels

This can be found in liver disease, pregnancy or as a result of excessive intravenous fluid infusion.

Measurement

(1) *Diacetyl monoxime methods.* In acid solution, diacetyl monoxime is converted to diacetyl which condenses with urea to form a coloured diazine derivative.

Ferric ions and thiosemicarbazide are included in reaction mixtures as these intensify the colour. Ammonia does not interfere with this reaction and therefore these methods can be used to estimate urine urea.

(2) *Urease–hypochlorite methods.* Urease hydrolyses urea to ammonia, which in the presence of the catalyst, nitroprusside, reacts with hypochlorite and phenol to give indophenol which is coloured blue in alkaline solution. In these methods care must be taken that all the water used for the reagents is ammonia free. In some methods salicylate is used instead of phenol.

(3) *Urease–glutamic dehydrogenase methods.* The ammonia liberated by the action of urease reacts with oxoglutarate to form glutamate, a reaction catalysed by the enzyme glutamate dehydrogenase. At the same time NADH is oxidized to NAD and this can be followed spectrophotometrically. This type of method can therefore be used on reaction rate analysers.

(4) *Urease–Nesslerization methods.* The ammonia formed by the urease reaction reacts with mercuric and potassium iodides in alkaline solution to give a yellow colour. However, this method has several disadvantages which has led to its gradual disuse. The major disadvantages are that:

(a) the reagents are not stable,
(b) Beer's law is not obeyed,
(c) protein-free filtrates must be used.

Further reading: General list of analytical and clinical textbooks.

UREA CLEARANCE

Urea is filtered at the glomerulus and about 40% is reabsorbed by passive diffusion from the renal tubules. Estimation of urea clearance therefore represents about 60% of the glomerular filtration rate. In most laboratories, however, creatinine clearance measurement is preferred to urea clearance.

URIC ACID

This is the end product of purine catabolism. It is excreted in the urine, probably as a result of active tubular secretion. Uric acid has only a limited solubility in the blood. Thus, in hyperuricaemic conditions, uric acid can be precipitated in tissues such as joints, skin and kidney, the latter leading to stone formation and renal failure. See also **gout.**

Causes of hyperuricaemia

(1) Primary gout. In this condition there is thought to be an increased synthesis of purines, possibly as a result of overactivity of one of the biosynthetic enzymes.

(2) Hyperuricaemia can occur in conditions where there is an increased turnover of nucleic acids, e.g. malignancies and tissue damage.

(3) As a result of reduced renal excretion of uric acid, e.g. renal disease.

(4) Lesch–Nyhan syndrome (juvenile hyperuricaemia). A rare inborn error of metabolism due to a deficiency of hypoxanthine-guanine phosphoribosyl transferase, an enzyme involved in the recycling of hypoxanthine and other purines.

Causes of hypouricaemia

(1) Hypouricaemia occurs in Fanconi syndrome.

(2) In the inborn error, xanthinuria, there is a deficiency of xanthine oxidase, an enzyme involved in the metabolism of purines to uric acid.

Measurement of serum uric acid

(1) Many methods of uric acid determination are based on the reduction of alkaline phosphotungstate or arsenotungstate by uric acid to tungsten blue. Sodium carbonate or cyanide is used as the alkali. A disadvantage of these methods is that other substances may also reduce phosphotungstate, giving falsely high values.

(2) Uric acid can be measured by its reduction of a metal complex, such as cupric phenanthroline, neocuproine or bathocuproine.

(3) Many methods are based on the enzyme, uricase. This catalyses the reaction:

$$\text{uric acid} + H_2O + O_2 \longrightarrow \text{allantoin} + CO_2 + H_2O_2$$

The enzymic reaction can be followed in a number of ways:

(a) By measuring the decrease in absorption of the uric acid at 290 nm.

(b) By measuring the oxygen consumed in the reaction by an oxygen electrode.

(c) By oxidation of a dye, e.g. *o*-dianisidine, by the hydrogen peroxide formed in the reaction, using the enzyme, peroxidase.

(d) By a modification of the Hantzsch reaction. The hydrogen peroxide formed in the enzymic reaction is used to oxidize methanol to formaldehyde. This reacts with diacetone alcohol and ammonia to give a yellow coloured complex.

(4) Uricase can also be used in conjunction with phosphotungstate or metal complex reduction. These methods are based on the decrease in the reducing substances brought about by the addition of uricase.

Detection of uric acid in stones

Uric acid can be detected in renal stones by the murexide test. This is the reaction of uric acid with nitric acid, followed by ammonium hydroxide to give ammonium purpurate (murexide) which is coloured purple.

Further reading: Balis, M.E. (1976). Uric acid metabolism in man. In Bodansky, O. and Latner, A.L. (eds.) *Advances in Clinical Chemistry*. Vol. 18. p. 213. (New York: Academic Press)

URINE ACIDIFICATION TEST

See: **ammonium chloride loading test**

URINE CONCENTRATION TEST

See: **water deprivation test**

URINE DILUTION TEST

See: **water excretion test**

URINE VOLUME

Normal urine volume in an adult is about 800–2000 ml/day. Increases in urinary volume (polyuria) occur in conditions such as hysterical polydipsia, diabetes insipidus, diabetes mellitus and Addison's disease. Low urine output occurs in dehydration, oedema and certain renal disorders.

URISTIX

A dipstick test, manufactured by Ames for the detection of protein and glucose in urine.

UROBILIN, UROBILINOGEN

Urobilinogen is a product of bilirubin metabolism, formed as a result of bacterial action in the gut (faecal urobilinogen is referred to as stercobilinogen). It is absorbed into the blood and the greater part is re-excreted by the liver while a small part is excreted in the urine where it can be oxidized to urobilin. Decreased faecal and urinary urobilinogen excretion occurs in biliary obstruction. Increased urinary excretion of urobilinogen occurs when there is liver cell damage, as in hepatitis.

Tests for the detection of urobilin or urobilinogen

(1) Schlesinger's test can be used for both urine and faeces. Iodine is used to convert urobilinogen into urobilin. Zinc acetate is added and the fluorescence of the zinc urobilin is assessed.

(2) Urine and faecal urobilinogen can be quantitated by its colorimetric reaction with Ehrlich's reagent.

UROBILISTIX

A dipstick test, manufactured by Ames for the semiquantitative estimation of urobilinogen in urine. It is based on the Ehrlich reaction.

Further reading: Kutter, D. (1977). *Rapid Clinical Diagnostic Tests.* (Munich–Vienna–Baltimore: Urban and Schwarzenberg)

UROCHROME

The pigment which gives urine its yellow colour.

UROERYTHRIN

The pink pigment, sometimes seen in urinary uric acid deposits with which it is precipitated.

UROPEPSIN, UROPEPSINOGEN

A small proportion of the pepsinogen secreted by the stomach finds its way into the bloodstream and is eventually excreted in the urine, where it is known as uropepsinogen. In the slightly acid pH of the urine, uropepsinogen is activated to uropepsin. Uropepsin output parallels gastric acid secretion and high excretion values are found in patients with duodenal ulcers. However, high outputs are also found in Cushing's syndrome and after physiological stress. Decreased output occurs in conditions associated with achlorhydria, e.g. pernicious anaemia.

Uropepsin can be measured by assaying tyrosine (using Folin–Ciocalteu reagent) which is liberated when haemoglobin is used as the substrate. A synthetic tyrosine substrate can also be used.

UROPORPHYRIN AND UROPORPHYRINOGEN

Uroporphyrinogen is an intermediate in the synthesis of haem. In certain types of porphyria there is increased conversion and excretion of uroporphyrins (porphyrins have acetate and propionate substituents). Elevated levels can be detected in faeces and urine by various screening tests (see **porphyrins**). If elevated urine levels of porphyrins are found, they can be identified as copro – or uroporphyrins by column or thin-layer chromatography. They can also be resolved by selective solvent extraction and estimated spectrophotometrically.

V

VALINE

A branched-chain amino acid found in increased levels in the blood, along with leucine and isoleucine, in the inborn error of metabolism, maple syrup urine disease. In the even rarer condition, hypervalinaemia, high serum valine levels are also found.

See also: **maple syrup urine disease.**

VALINE LOADING TEST

A test which can be used in the investigation of vitamin B_{12} deficiency. The metabolism of valine contains a step in which methylmalonyl coenzyme A is converted to succinyl coenzyme A, an enzymic reaction which requires vitamin B_{12} as a cofactor. In vitamin B_{12} deficient individuals, administration of an oral load of valine results in a greater than normal excretion of methylmalonate in the urine which can be estimated colorimetrically.

See also: **vitamin B_{12}**

VAN DEN BERGH REACTION

The reaction of bilirubin with diazotized sulphanilic acid to form azobilirubin. It is widely used for the estimation of bilirubin.

See: **bilirubin**

VANILLYL MANDELIC ACID

See: **4-hydroxy-3-methoxymandelic acid**

VARIANCE

A measure of the range of the observations in a sample. It can be found by calculating the difference between each observation (x)

and the mean value of the observations ($\bar{x}$). These differences are squared. The mean value of these differences

$$\text{variance} = \frac{\Sigma(x - \bar{x})^2}{n - 1}$$

where n = the number of observations

is called the variance. However, as it is desirable to have a measure of spread in the same units as the original observations, the square root of the variance is often calculated. This is the standard deviation (q.v.).

VASOACTIVE INTESTINAL PEPTIDE (VIP)

This compound is synthesized by the mucosal endocrine cells in the gut. VIP-producing tumours are known and these can be diagnosed by finding high plasma levels of VIP.

VASOPRESSIN

See: **antidiuretic hormone**

VASOPRESSIN TEST

See: **water deprivation test**

VERY LOW DENSITY LIPOPROTEINS

Synonym for pre-β-lipoproteins.

See: **lipoproteins**

VITAMIN A (RETINOL)

A fat-soluble vitamin, most of which is obtained from its precursor molecule, β-carotene, found in plants. This is hydrolysed in the gut to give vitamin A. The vitamin is required for mucopolysaccharide synthesis and mucus secretion. It is also combined with the protein, opsin, in the retinal pigment, rhodopsin, which is necessary for vision in the dim light. A deficiency of the vitamin causes night blindness and drying of the cornea and conjunctivae. Blood vitamin A levels can be measured in order to diagnose deficiency conditions.

Measurement

Several methods are available:

(1) Vitamin A can be extracted from serum into organic solvents. The extract is divided into two and one of the portions is irradiated with ultraviolet light which destroys the vitamin A. The two portions are then read at 327 nm, the difference in absorbance being due to the vitamin A.

(2) Trifluoroacetic acid reacts with vitamin A to give a blue colour (the Neeld–Pearson procedure).

(3) Vitamin A reacts with antimony trichloride in chloroform to give a blue colour (the Carr–Price reaction).

Further reading: General list of analytical and clinical textbooks

VITAMIN A ABSORPTION TEST

A test used for the diagnosis of steatorrhoea. It consists of giving oral vitamin A and measuring the rise in serum vitamin A levels. Diminished absorption occurs in steatorrhoea.

VITAMIN B COMPLEX

The water-soluble vitamin B complex consists of thiamine (vitamin B_1), riboflavin (vitamin B_2), nicotinamide, pyridoxine (vitamin B_6), biotin, pantothenic acid, folic acid and vitamin B_{12}. See separate entries for these compounds.

VITAMIN B_{12} (CYANOCOBALAMIN, EXTRINSIC FACTOR)

A cobalt-containing vitamin which is required for normal haemopoiesis. It participates as a cofactor in certain enzymic reactions, including the synthesis of methionine from homocysteine, the conversion of methyl malonyl CoA to succinyl CoA, and the methylation of RNA. Its role as a cofactor in the synthesis of nucleic acids is closely liked to that of folate.

Absorption

Intrinsic factor is required for the absorption of vitamin B_{12}. This is a glycoprotein secreted by the mucosal cells of the stomach which combines with the vitamin and is then transported into the mucosal cells of the ileum.

Deficiency

Vitamin B_{12} is obtained from the diet in milk, eggs and liver. A dietary deficiency of the vitamin is rare. A deficiency may occur when there is reduced absorption in diseases which affect the terminal ileum such as Crohn's disease. The amount of vitamin B_{12} available for absorption is also reduced when there is increased bacterial colonization of the small intestine. The most important cause of vitamin B_{12} deficiency is, however, a failure to secrete intrinsic factor, a condition which may have an autoimmune basis, since intrinsic factor antibodies can be demonstrated in affected individuals.

Symptoms of vitamin B_{12} deficiency

Vitamin B_{12} deficiency results in pernicious anaemia. In this condition there is a delayed maturation of erythrocytes due to the impairment of DNA synthesis. This results in the appearance of megaloblasts in the blood. Deficiency of vitamin B_{12} also results in a neurological condition (subacute combined degeneration of the spinal cord).

Assessment of vitamin B_{12} deficiency

(1) By measurement of the serum levels of the vitamin. This can be achieved by competitive protein binding (using, for example, intrinsic factor as the binding protein) or by biological assay. The biological assay is performed by measuring the growth, in the presence of the patients serum, of vitamin B_{12}-requiring strains of micro-organism. Such bioassays are now little used however.

(2) By measuring the urinary excretion of methylmalonic acid. This is an intermediate in the metabolism of certain amino acids, particularly valine and isoleucine. Vitamin B_{12} is a cofactor in the step in which methylmalonyl CoA is converted to succinyl CoA. Increased urinary excretion of methylmalonate therefore occurs in vitamin B_{12} deficient states. Methylmalonate excretion can also be measured after a loading dose of valine (see **valine loading test**).

(3) By measurement of the absorption of ^{57}Co- or ^{58}Co-labelled vitamin B_{12} (the Schilling test). Oral doses of the labelled vitamin are given and its subsequent absorption and excretion in the urine is measured. If abnormal results are obtained, labelled vitamin together with intrinsic factor

can be given. If normal results are now obtained, it suggests that the patient is not producing enough of his own intrinsic factor. Commercial kits are available which enable both tests to be performed together.

Further reading: Chanarin, I. (1977). Folates, cobalamins and their interrelationship in man. In Marks, V. and Hales, C.N. (eds.) *Essays in Medical Biochemistry*. Vol. 3, p. 1. (London: The Biochemical Society and the Association of Clinical Biochemists)

VITAMIN C

See: **ascorbic acid**

VITAMIN D

A group of fat-soluble vitamins. Cholecalciferol (vitamin D_3) can be obtained from dietary sources and can also be produced in the skin by the action of sunlight on 7-dehydrocholesterol. Another vitamin D (ergocalciferol, vitamin D_2) can be obtained from plants. Cholecalciferol undergoes two hydroxylation steps in the body in its conversion to its active form, 1,25-dihydroxycholecalciferol. The C-25 hydroxylation takes place in the liver and this is followed by the C-1 hydroxylation which takes place in the kidneys. The rate of synthesis of 1,25-dihydroxycholecalciferol is thought to be controlled by the circulating parathyroid hormone levels.

Actions

1,25-dihydroxycholecalciferol increases calcium absorption in the intestine. Together with parathyroid hormone, it stimulates calcium release from bone. It also increases phosphate excretion in the urine.

Deficiency

A deficiency of vitamin D results in rickets in children and osteomalacia in adults. It can occur as a result of malnutrition, or malabsorption. Deficiency can also occur when there is a failure to hydroxylate vitamin D. This is thought to occur in some types of renal disease and as a result of treatment with particular anticonvulsant drugs. These drugs are thought to stimulate liver enzymes which oxidize vitamin D to inactive metabolites.

Overdosage

Overdosage with vitamin D can cause hypercalcaemia. Oversensitivity to vitamin D is thought to be the cause of two hypercalcaemic conditions, sarcoidosis and idiopathic hypercalcaemia of infancy.

Measurement

Vitamin D can be measured by biological or immunological methods.

Further reading: DeLuca, H.F. (1977). Vitamin D endocrine system. In Bodansky, O and Latner, A.L. (eds.) *Advances in Clinical Chemistry*. Vol. 19, p. 125. (New York: Academic Press)

VITAMIN E (TOCOPHEROLS)

Deficiency of this compound does not appear to have any clinical effects in humans.

VITAMIN K

A fat-soluble vitamin required for prothrombin synthesis. Deficiency of the vitamin results in a bleeding tendency.

VMA

See: **4-hydroxy-3-methoxymandelic acid**

VON GIERKE'S DISEASE

The commonest form of glycogen storage disease (qv).

W

WALDENSTRÖMS MACROGLOBULINAEMIA

A condition in which there is a malignant proliferation of cells resembling lymphocytes, resulting in the production of a monoclonal IgM paraprotein. The presence of large amounts of this high molecular weight protein in the plasma leads to sluggish blood flow, and, as a result, thrombosis of the small blood vessels can occur ('hyperviscosity syndrome'). Retinal vein thrombosis, cerebral thrombosis and peripheral gangrene are possible consequences. The condition can be treated by exchange plasma transfusions.

Further reading: General list of clinical textbooks

WATER BALANCE

The water content of an average 70 kg man is approximately 45 litres and this is maintained by a balance between intake and loss.

Water intake

The body obtains water from food and drink. The hypothalamic thirst control centre is involved in the regulation of fluid intake.

Water loss

(1) Nearly one litre of water is lost per day from the lungs and in sweat (insensible loss).

(2) Some fluid is lost in the faeces.

(3) Fluid is also lost through the kidneys, a process which is regulated by antidiuretic hormone (ADH). Secretion of large amounts of ADH results in more water reabsorption by the renal tubules and the formation of a concentrated urine. Decreased secretion of ADH results in less water reabsorption and a dilute urine being produced.

Disturbances of water balance

The homeostatic mechanisms of water balance and sodium balance are closely linked. Nevertheless it is possible to identify particular conditions where the primary disorder is one of abnormal fluid balance rather than abnormal sodium balance.

Conditions in which there is primary water depletion

(1) When there is excessive loss of fluid in gastro-intestinal secretions, sweat, lungs or in the urine the latter as a result of ADH deficiency (diabetes insipidus).

(2) When there is deficient water intake as in thirst or comatose patients.

Hypernatraemia is a feature of primary water depletion, although this must be distinguished from other causes of hypernatraemia due to disorders of sodium metabolism.

Conditions in which there is primary water excess

(1) In 'inappropriate' ADH secretion by tumours or other conditions. This results in excessive water reabsorption by the renal tubules.

(2) In renal failure, due to incorrect fluid therapy.

Hyponatraemia is a feature of primary water excess, although it must be distinguished from other causes of hyponatraemia due to disorders of sodium metabolism.

See also: **antidiuretic hormone**

Further reading: General list of clinical textbooks

WATER DEPRIVATION TEST (URINE CONCENTRATION TEST)

A procedure which tests the ability of the kidney to produce a concentrated urine. By depriving an individual of fluid intake, ADH secretion should be maximal and a concentrated urine should be produced. If there is a failure in the ability to secrete ADH (as in diabetes insipidus) or if there is a renal disorder, no concentration of the urine occurs (as measured by specific gravity or osmolality). If abnormal results are obtained, the test can be repeated and, at the same time, pitressin (ADH) can be injected. A further failure to concentrate the urine is indicative of renal disease rather than failure to secrete ADH.

Further reading: Mitchell, F.L., Veall, N. and Watts, R.W.E. (1972). Scientific Review No. 2. Renal function tests suitable for clinical practice. *Ann. Clin. Biochem.,* **9**, 1

WATER EXCRETION TEST (URINE DILUTION TEST)

A procedure which tests the ability of the kidney to excrete a water load. A failure to excrete the major part of a water load could indicate renal disease, oedema or adrenal insufficiency.

Further reading: Mitchell, F.L., Veall, N. and Watts, R.W.E. (1972). Scientific Review No. 2. Renal function tests suitable for clinical practice. *Ann. Clin. Biochem.,* **9**, 1

WILSON'S DISEASE (HEPATOLENTICULAR DEGENERATION)

An inherited disease in which there are low levels of caeruloplasmin in the serum. As a consequence of this, copper is deposited in the tissues, particularly in the basal ganglia of the brain (leading to neurological symptoms), in the liver (leading to cirrhosis), in the renal tubules (leading to renal tubular damage and aminoaciduria), and in the eyes (giving Kayser–Fleischer rings). The condition can be diagnosed biochemically by demonstrating low plasma caeruloplasmin and copper levels and increased urinary excretion of copper. The disease can be treated by compounds such as penicillamine which chelate copper and reduce the tissue copper concentration.

Further reading: Sass-Kortsak, A. and Bearne, A.C. (1978). Hereditary disorders of copper metabolism. In Stanbury, J.B., Wyngaarden, J.B. and Fredrickson, D.S. (eds.) *The Metabolic Basis of Inherited Disease*. 4th Edn., p. 1098. (New York: McGraw-Hill)

WOHLGEMUTH UNIT (DIASTATIC INDEX)

A unit in which amylase activity can be expressed.

WOLMAN'S DISEASE

A rare inborn error of metabolism in which large amounts of cholesterol esters and triglycerides are deposited in the tissues with fatal consequences.

Further reading: Fredrickson, D.S. and Ferrans, V.J. (1978). Acid cholesteryl ester hydrolase deficiency. (Wolman's disease and cholesteryl ester storage disease). In Stanbury, J.B., Wyngaarden, J.B. and Fredrickson, D.S. (eds.) *The Metabolic Basis of Inherited Disease*. 4th Edn., p. 670. (New York: McGraw-Hill)

X

XANTHINE

An intermediate in the formation of uric acid from purines. It is excreted in the urine in large amounts in the inborn error of metabolism, xanthinuria, a condition in which xanthine stones can be found. It can be detected in urinary stones by its reaction with Ehrlich's diazo reagent to give a red colour.

XANTHINE OXIDASE

An enzyme involved in the formation of uric acid from purines. Drugs, such as allopurinol, inhibit xanthine oxidase and are used in the treatment of gout. A deficiency of the enzyme occurs in the inborn error of metabolism, xanthinuria.

XANTHINURIA

A rare inborn error of metabolism in which there is a deficiency of xanthine oxidase, an enzyme involved in the conversion of purines to uric acid. As a result large amounts of xanthine are excreted in the urine and this can result in the formation of xanthine stones. The mode of inheritance is probably that of an autosomal recessive.

Further reading: Wyngaarden, J.B. (1978). Hereditary xanthinuria. In Stanbury, J.B., Wyngaarden, J.B. and Fredrickson, D.S. (eds.) *The Metabolic Basis of Inherited Disease*. 4th Edn., p. 1037. (New York: McGraw-Hill)

XANTHOCHROMIA

The presence of a yellow colour in cerebrospinal fluid. It can be due to degraded haemoglobin after a cerebral haemorrhage or can occur as a result of jaundice.

XANTHOMATA

Yellow deposits of lipids in tissues of the body, including the skin. They can be found in hyperlipidaemic conditions.

XANTHURENIC ACID

This is the excretion product of 3-hydroxykynurenic acid, an intermediate in the conversion of tryptophan to nicotinic acid. Pyridoxal phosphate is required as a cofactor for the enzyme, kynureninase, which catalyses the conversion of 3-hydroxykynurenic acid to 3-hydroxyanthrinilic acid. In patients with pyridoxine deficiency, 3-hydroxykynurenic acid accumulates and is excreted in the urine as xanthurenic acid. Xanthurenic acid can therefore be measured in urine (especially after giving an oral typtophan load) in order to detect pyridoxine deficiency.

Measurement

(1) Colorimetric methods based on the formation of a green colour with ferric salts in sodium bicarbonate solution.

(2) Xanthurenic acid can be estimated by its fluorescence after its isolation by ion-exchange chromatography.

See also: **pyridoxine**

X-RAY FILM TEST

A test for the semi-quantitative measurement of trypsin activity in faeces or duodenal fluid. It consists of placing serial dilutions of the sample on unexposed X-ray film. Tryptic activity results in the formation of clear areas on the film as a result of enzymic hydrolysis of the gelatin layer. The greatest dilution of the specimen giving a clear area is an approximate indication of the amount of trypsin present.

XYLOSE ABSORPTION TEST

D-Xylose is a pentose not normally found in the blood. When an oral load is given to the patient, it is passively absorbed in the proximal small intestine and eventually excreted unchanged in the urine. Decreased absorption of xylose occurs in intestinal malabsorption providing this is not due to pancreatic insufficiency.

Measurement of xylose

Xylose, when heated with acids, forms furfural which in turn reacts with *p*-bromoaniline to produce a pink colour.

Further reading: General list of analytical and clinical textbooks

Z

ZIMMERMANN REACTION

The reaction of 17-oxosteroids with *m*-dinitrobenzene in strong alkali to produce a red-violet coloured compound. This reaction is used for the quantitative estimation of 17-oxosteroids and also for 17-hydroxysteroids after their chemical conversion to 17-oxosteroids.

See also: **17-oxogenic steroids, 17-oxosteroids**

ZINC

A metal required in trace amounts. It is a cofactor in certain enzymic reactions and it is thought to promote wound healing. It can be measured in biological fluids by atomic absorption spectrophotometry.

Further reading: Mikac-Devic, D. (1970). Methodology of zinc determinations and the role of zinc in biochemical processes. In Bodansky, O. and Stewart, C.P. (eds.) *Advances in Clinical Chemistry*. Vol. 13, p. 271. (New York: Academic Press)
Editorial. (1975). Zinc in human medicine. *Lancet*, **2**, 351

ZINC SULPHATE TURBIDITY

A flocculation test for the detection of increased levels of γ-globulins in serum. It consists of addition of the serum sample to zinc sulphate solution and assessing the degree of turbidity, which is proportional to the increase in the γ-globulin fraction.

ZOLLINGER-ELLISON SYNDROME

This is a condition in which there is excessive secretion of gastric juice as a result of gastrin secretion, usually by a pancreatic tumour or, less commonly, from hyperplasia of the G cells of the stomach. The continuous secretion of gastric juice results in severe peptic ulceration and possibly steatorrhoea. The estimation of serum gastrin can assist in the diagnosis.

Further reading: General list of clinical textbooks

ZONE ELECTROPHORESIS

The separation of a mixture of substances into zones on a supporting structure, e.g. cellulose acetate, paper, starch gel or agar.

General List of Analytical Textbooks

1. Bender, G. T. (1972). *Chemical Instrumentation: a Laboratory Manual Based on Clinical Chemistry.* (Philadelphia, London, Toronto: W. B. Saunders Co.)
2. Henry, R. J., Cannon, D. C. and Winkelman, J. W. (eds.) (1974) *Clinical Chemistry. Principles and Technics.* 2nd Edn. (Hagerstown, Maryland: Harper and Row)
3. Kamath, S. H. (1972) *Clinical Biochemistry for Medical Technologists.* (Edinburgh: Churchill Livingstone)
4. Kaplan, A. and Szabo, L. L. (1979) *Clinical Chemistry: Interpretation and Techniques.* (Philadelphia: Lea and Febiger)
5. Tietz, N. W. (ed.) (1976). *Fundamentals of Clinical Chemistry.* 2nd Edn. (Philadelphia, London, Toronto: W. B. Saunders Co.)
6. Toro, G. and Ackermann, P. G. (1975) *Practical Clinical Chemistry.* (Boston: Little, Brown)
7. Williams, D. L., Nunn, R. F. and Marks, V. (eds.) (1978). *Scientific Foundations of Clinical Biochemistry.* Vol. 1. Analytical Aspects. (London: Heinemann)
8. Wolf, P. L., Williams, D., Tsudaka, T. and Acosta, L. (1972). *Methods and Techniques in Clinical Chemistry.* (Chichester: John Wiley)
9. Wooton, I. D. P. (1974). *Microanalysis in Medical Biochemistry.* 5th Edn. (Edinburgh: Churchill Livingstone)
10. Varley, H., Gowenlock, A. H. and Bell, M. (1979). *Practical Clinical Chemistry.* 5th Edn. Vol. 1. (In press.) Vol. 2. 1976. Hormones, Vitamins, Drugs and Poisons. (London: Heinemann)

General List of Clinical Textbooks

1. Baron, D. N. (1973) *A Short Textbook of Chemical Pathology* 3rd Edn. (London: English Universities Press)
2. Gray, C. H. and Howorth, P. J. N. (1977) *Clinical Chemical Pathology*. 8th Edn. (London: Edward Arnold)
3. Hyde, T. A. and Draisey, T. F. (1974). *Principles of Chemical Pathology*. (London: Butterworths). (Also contains some analytical details)
4. Latner, A. L. (1975). *Cantarow and Trumper. Clinical Biochemistry*. 7th Edn. (Philadelphia, London, Toronto: W. B. Saunders Co.)
5. Whitby, L. G., Percy-Robb, I. W. and Smith, A. F. (1975). *Lecture Notes on Clinical Chemistry*. (Oxford: Blackwell Scientific Publications).
6. Widmann, F. K. (1979). *Clinical Interpretation of Laboratory Tests*. (Philadelphia: F. A. Davis Co.)
7. Zilva, J. F. and Pannall, P. R. (1979). *Clinical Chemistry in Diagnosis and Treatment*. 3rd Edn. (London: Lloyd-Luke).

"Normal" ranges for some of the more commonly measured constituents in biological fluids

Enzymes are not included as the normal range varies considerably according to the method used

		SI Units	Other Units
Serum (S), Plasma (P), or Blood (B).			
S	Albumin	35–47 g/l	3.5–4.7 g/100 ml
P	Ammonium (Ion)	37–84 μmol/l	68–153 μg/100 ml
B	Ascorbate	45–80 μmol/l	0.8–1.4 mg/100 ml
B	Base Excess	±2 mmol/l	±2 mEq/l
P	Bicarbonate	24–32 mmol/l	24–32 mEq/l
P	Bilirubin (Total)	3–17 μmol/l	0.2–1.0 mg/100 ml
P	Bilirubin (Direct)	0–3 μmol/l	0–0.2 mg/100 ml
S	Caeruloplasmin	300–600 mg/l	30–60 mg/100 ml
S	Calcium	2.1–2.6 mmol/l	8.5–10.5 mg/100 ml
S	Carotene	1.1–3.7 μmol/l	60–200 μg/100 ml
S	Chloride	98–108 mmol/l	98–108 mEq/l
S	Cholesterol	3.6–7.8 mmol/l Varies with age	140–300 mg/100 ml Varies with age
B	PCO_2	4.7–6.1 kPa	35–46 mmHg
P	Copper	13–24 μmol/l	80–150 μg/100 ml
P	Cortisol (9–10 a.m.)	180–730 nmol/l	6.5–26.0 μg/100 ml
S	Creatinine	53–106 μmol/l	0.6–1.2 mg/100 ml
P	Fibrinogen	1.5–4.0 g/l	150–400 mg/100 ml
S	Folate	3–20 μg/l	3–20 ng/ml
P	Glucose (Fasting)	3.9–5.8 mmol/l	70–105 mg/100 ml
S	Iron	13–32 μmol/l	70–180 μg/100 ml
S	Iron Binding Capacity (Total)	45–70 μmol/l	250–400 μg/100 ml
B	Ketones (As Acetoacetate)	80–140 μmol/l	0.8–1.4 mg/100 ml
B	Lactate (Arterial)	0.3–0.8 mmol/l	3.1–7.0 mg/100 ml
B	Lead	0.5–2.0 μmol/l	10–40 μg/100 ml
P	Lipids (Total)	4.0–10.0 g/l	400–1000 mg/100 ml

		SI Units	Other Units
P	Magnesium	0.7–1.0 mmol/l	1.8–2.4 mg/100 ml
B	PO_2	12–15 kPa	90–110 mmHg
P	Osmolality	275–295 mmol/kg	275–295 mosmol/kg
B	pH	7.35–7.45	7.35–7.45
P	Phenylalanine	0.06–0.2 mmol/l	1.00–3.0 mg/100 ml
P	Phosphate (Inorganic)	0.8–1.4 mmol/l	2.5–4.5 mg/100 ml
P	Potassium	3.8–5.0 mmol/l	3.8–5.0 mEq/l
S	Protein (Total)	62–82 g/l	6.2–8.2 g/100 ml
B	Pyruvate	58–196 μmol/l	0.5–1.7 mg/100 ml
P	Sodium	136–148 mmol/l	136–148 mEq/l
S	Thyroxine	70–160 mmol/l	5.5–12.5 μg/100 ml
S	Transferrin	1.2–2.0 g/l	120–200 mg/100 ml
P	Triglyceride (As Triolein)	0.3–1.7 mmol/l	25–150 mg/100 ml
P	Urate	0.1–0.4 mmol/l	2–7 mg/100 ml
B	Urea	2.5–6.5 mmol/l	15–40 mg/100 ml
P	Vitamin A	0.7–1.7 μmol/l	20–50 μg/100 ml
S	Vitamin B_{12} (As Cyanocobalamin)	160–925 ng/l	160–925 pg/ml

Urine

	SI Units	Other Units
Ascorbate	110–280 μmol/24 h	20–50 mg/24 h
Calcium	2.5–7.5 mmol/24 h	100–300 mg/24 h
Catecholamines (As Adrenaline)	0.05–0.55 μmol/24 h	10–100 μg/24 h
Copper	0.2–0.8 μmol/24 h	10–50 μg/24 h
Coproporphyrins	0.15–0.31 μmol/24 h	100–200 μg/24 h
Creatine	0–400 μmol/24 h	0–50 mg/24 h
Creatinine	9–17 mmol/24 h	1.0–2.0 g/24 h
Formimino-glutamate	0–170 μmol/24 h	0–30 mg/24 h
5-Hydroxyindole-acetic Acid	15–75 μmol/24 h	3–14 mg/24 h
4-Hydroxy-3-meth-oxymandelic Acid	10–35 μmol/24 h	2–7 mg/24 h
Hydroxyproline	0.08–0.25 mmol/24 h	10–35 mg/24 h
Lead	0.14–0.40 μmol/24 h	30–80 μg/24 h
Magnesium	3.3–5.0 mmol/24 h	80–120 mg/24 h
Oxalate	0.2–0.4 mmol/24 h	20–40 mg/24 h
17-Oxogenic	35–70 μmol/24 h	10–20 mg/24 h
Steroids	Varies with age and sex	

	SI Units	Other Units
17-Oxosteroids	35–85 μmol/24 h	10–25 mg/24 h
	Varies with age and sex	
Phosphate (Inorganic)	15–50 mmol/24 h	0.5–1.5 g/24 h
Porphobilinogen	1–10 μmol/24 h	0.2–2.0 mg/24 h
Potassium	25–100 mmol/24 h	25–100 mEq/24 h
Sodium	100–200 mmol/24 h	100–200 mEq/24 h
Urate	3–12 mmol/24 h	0.5–2.0 g/24 h
Urea	250–500 mmol/l	1.5–3.0 g/100 ml
Uroporphyrins	0–30 nmol/24 h	0–25 μg/24 h

Faeces

Fat (As Stearic Acid)	11–18 mmol/24 h	3–5 g/24 h
Urobilinogen	50–500 μmol/24 h	30–300 mg/24 h

GPSR Compliance

The European Union's (EU) General Product Safety Regulation (GPSR) is a set of rules that requires consumer products to be safe and our obligations to ensure this.

If you have any concerns about our products, you can contact us on ProductSafety@springernature.com

In case Publisher is established outside the EU, the EU authorized representative is:

Springer Nature Customer Service Center GmbH
Europaplatz 3
69115 Heidelberg, Germany

Batch number: 10584446

Printed by Printforce, the Netherlands